T0281100

Cambridge Elements ☰

Elements in the Philosophy of Biology
edited by
Grant Ramsey
KU Leuven
Michael Ruse
Florida State University

THE ROLE OF MATHEMATICS IN EVOLUTIONARY THEORY

Jun Otsuka
Kyoto University

CAMBRIDGE
UNIVERSITY PRESS

CAMBRIDGE
UNIVERSITY PRESS

University Printing House, Cambridge CB2 8BS, United Kingdom

One Liberty Plaza, 20th Floor, New York, NY 10006, USA

477 Williamstown Road, Port Melbourne, VIC 3207, Australia

314–321, 3rd Floor, Plot 3, Splendor Forum, Jasola District Centre,
New Delhi – 110025, India

79 Anson Road, #06–04/06, Singapore 079906

Cambridge University Press is part of the University of Cambridge.

It furthers the University's mission by disseminating knowledge in the pursuit of
education, learning, and research at the highest international levels of excellence.

www.cambridge.org
Information on this title: www.cambridge.org/9781108727853
DOI: 10.1017/9781108672115

First published 2019

A catalogue record for this publication is available from the British Library.

ISBN 978-1-108-72785-3 Paperback
ISSN 2515-1126 (online)
ISSN 2515-1118 (print)

The Role of Mathematics in Evolutionary Theory

Elements in the Philosophy of Biology

DOI: 10.1017/9781108672115
First published online: October 2019

Jun Otsuka
Kyoto University

Abstract: The central role of mathematical modeling in modern evolutionary theory has raised a concern as to why and how abstract formulae can say anything about empirical phenomena of evolution. This Element introduces existing philosophical approaches to this problem and proposes a new account according to which evolutionary models are based on causal, and not just mathematical, assumptions. The novel account features causal models both as the Humean "uniform nature" underlying evolutionary induction and as the organizing framework that integrates mathematical and empirical assumptions into a cohesive network of beliefs that functions together to achieve epistemic goals of evolutionary biology.

Keywords: evolutionary theory, causal modelling, fitness, mathematical explanation, population genetics

ISBNs: 9781108727853 (PB), 9781108672115 (OC)
ISSNs: 2515-1126 (online), 2515-1118 (print)

Contents

1 Math for Evolution: Holy Grail or Poisoned Chalice?

Like any other advanced science, contemporary evolutionary theory is highly mathematized. The history and dynamics of evolutionary processes are described and explained in the language of probability, differential equations, and linear algebra, as can be easily confirmed by a look at major journals like *Evolution, Genetics*, or *The American Naturalist* or standard textbooks on evolutionary genetics. This is in stark contrast with Darwin's *Origin of Species*, which established the fact and principle of evolution with an overwhelming mass of empirical examples but not a single equation. Evolutionary theory, therefore, was mathematized at some point after its birth – but when, and why?

Although mathematics surely did not give birth to Darwin's theory, it *saved* its life, or at least the life of one of its halves – halves, because Darwin submitted two logically independent theses in his *Origin*. One is the historical hypothesis that diverse forms of life on earth have emerged by the branching of a few or possibly just one primitive kind. This is called *the principle of common descent*, the one that tells you that humans are distant relatives of bacteria. Darwin's second thesis, which was also independently proposed by Alfred Wallace, is the famous *principle of natural selection*, which claims that evolutionary changes and speciation occur because individuals in a population differ in their ability to survive and reproduce, and these abilities tend to be inherited by their offspring.

Of these two theses, the principle of common descent was soon accepted with little antagonism, at least in the scholarly circle of the late nineteenth century. Many biologists, however, resisted the idea of natural selection as a major cause of such historical changes for three reasons (Provine, 2001). The first source of disagreement was insufficient knowledge about the mechanism of inheritance. Darwin presupposed a sort of "blending" inheritance, whereby parents' characteristics mix in their offspring's phenotype so that if a new mutant with an advantageous characteristic mates with an average individual in the population, their offspring will show an intermediate character. But this entails that any advantageous trait that arises with a single mutant in a population will be diluted away after a few rounds of sexual reproduction, well before it could be spread throughout the population or species by selection (Fig. 1.1). Selection would then require a large number of mutational inputs to change the population structure, making it no more than a negligible factor. The second issue came from failures of contemporary experiments that tried to create a new species or a significant variant with visible morphological differences by repeated artificial selection in the laboratory. The negative results of these experiments suggested the existence of a deep gap between species that natural selection cannot overcome.

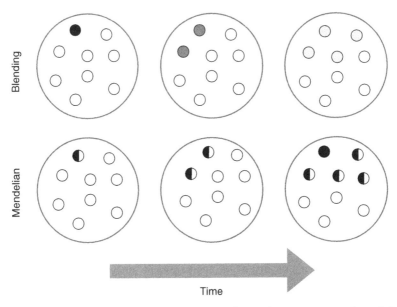

Time

Figure 1.1 If parents traits "blended" in offspring's phenotype, a rare beneficial characteristic would be diluted away before selection can act on it. In Mendelian inheritance, a parent's allele is passed on to offspring as it is.

Sympathizers of Darwin could and did respond that adaptive processes take a much longer time than the duration of these experiments; but having no means to confirm consequences of such lengthy processes, their rebuttal remained speculative and held no water against most contemporary critics with a positivist slant who, in the wake of experimental biology, put much emphasis on well-controlled and reproducible experiments. The third obstacle for the Darwinian theory concerned its compatibility with Mendelian genetics, which was rediscovered around the beginning of the twentieth century and soon became accepted as a correct description of the mechanism of inheritance. Mendel's pea experiment showed that organismal characteristics "jump" from one type to another (yellow or green, smooth or wrinkled) by the transmission of discrete factors we now call genes or alleles. This result, however, seemed to contradict Darwin's claim that evolution by natural selection is a gradual process that acts on subtle, and mostly continuous, variations. Moreover, such a gradual evolution appeared to be far less effective. Upon the observation that significant morphological changes are often triggered by single genetic replacements, Mendelian geneticists concluded that the creation of new variants or "sports" by such mutations plays a far more important role in major evolutionary changes and speciation than does selection. Faced with these criticisms and difficulties, the Darwinian theory of natural selection around the turn of the twentieth century was almost abandoned, even to the extent that American biologist Vernon Kellogg worried that Darwinism was on its "death-bed" (Bowler, 1983).

As seen from the discussion above, the skepticism against natural selection was targeted at its *efficacy*. The suspicion was not that natural selection is impossible, but that it plays only a minor and secondary role compared to other evolutionary forces such as mutation. To this question of degree, Darwin's qualitative and schematic arguments in his *Origin* had no substantive answer. The resolution of these issues had to wait for the integration of Darwin's theory with the Mendelian theory of genetics, in which mathematical formulations of selection and reproduction played an essential role in showing how and to what extent selection can alter biological populations (Provine, 2001, ch. 5). On the first problem of blending inheritance, G. H. Hardy and W. Weinberg independently found in 1908 that in the absence of other evolutionary forces (such as selection or migration) the genotype frequency of a Mendelian population stabilizes at a fixed ratio we now call the Hardy–Weinberg equilibrium. If, for instance, there are two alleles A, a in the population with the frequencies $Pr(A) = p$ and $Pr(a) = q = 1 - p$, the relative frequencies of the three genotypes $AA{:}Aa{:}aa$ stay $p^2 : 2pq : q^2$. This implies that a variation introduced into a population as a few mutant alleles will not get diluted away but will remain as it is, giving room and opportunity for selection to increase its frequency. But how long does this process take? To examine the second skeptical argument that selection alone cannot achieve much evolutionary change, R. C. Punnett and his fellow mathematician H. T. J. Norton calculated the number of generations required for selection to change gene frequency in a population (Punnett, 1915). The result of their numerical calculation showed that even a trait with the slightest selective advantage can sweep through a population in a relatively short period, vindicating the efficacy of gradual selection. These mathematical developments culminated in Ronald Fisher's (1918) formal integration of the Darwinian theory of selection with Mendelian genetics, which reduced gradual evolution of continuous traits (such as height) to frequency changes of a large array of underlying genes, each having a small phenotypic effect. This result allowed Fisher to calculate and predict the evolutionary response of a continuous phenotype to a given selective pressure (see Section 5) and to reformulate the Darwinian gradual evolution within the Mendelian framework, showing their logical consistency. These theoretical developments dispelled the skepticism against the Darwinian theory, and in the early 1920s natural selection came to be acknowledged as one of the most important forces to produce evolutionary change.

The formal integration with Mendelian genetics not only helped Darwin's theory of natural selection, but also put mathematics at the center of evolutionary studies. Fisher's work, along with other seminal contributions by S. Wright and J. B. S. Haldane, gave birth to the new field of *population genetics* and formed the theoretical core of the "Modern Synthesis," the standard paradigm of evolutionary studies in the twentieth century. The canonical evolutionary theory characterizes organisms by two aspects, *genotype* (a set of genes organisms possess) and *phenotype* (any other physiological, morphological, or behavioral features such as height or metabolic rate), and accordingly a population of organisms by its genetic and phenotypic distributions. Population distributions can be formally represented as points in *genetic or phenotypic spaces*, in which evolutionary processes are conceptualized as trajectories or movements of these points (Lloyd, 1988). Since genotypic and phenotypic characterizations are two sides of the same coin, evolution in each space does not proceed independently but rather runs side-by-side. Lewontin (1974, see also Fig. 1.2) illustrates this tandem evolution as consisting of four transitional steps, namely: (T_1) development from fertilized eggs/genotype into adult form/phenotype; (T_2) change in phenotypic distributions due to selection,

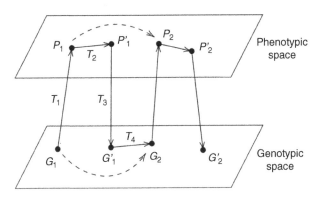

Figure 1.2 Schematic representation of evolutionary processes adopted from Lewontin (1974). G and P are respectively genotypic and phenotypic descriptions of an evolving population, with $T_1 \sim T_4$ denoting steps in the life cycle. (T_1) A population of fertilized eggs (zygotes) G_1 develops into a population of adult individuals P_1. (T_2) Selection and other evolutionary forces act to alter the population composition. (T_3) Surviving individuals P_1' create eggs and sperm (gametes) G_1'. (T_4) Gametes combine and form the zygotes G_2 of the next generation, and the process continues. Although phenotypic and genotypic evolution thus proceed in tandem, most models focus on tracking changes in one dimension, as indicated by the dashed arrows.

migration, etc.; (T_3) gamete (eggs and sperm) production by surviving individuals; and (T_4) fertilization and formation of new genotypes. The "horizontal" transitions T_2 and T_4 represent shifts in phenotypic or genotypic distribution of a population by various evolutionary factors, while the "vertical" transitions T_1 and T_3 transcribe back and forth between the phenotypic and genotypic aspects of the population. The goal of population genetics is to build a mathematical model that takes into account all these transitional steps so that it accurately tracks the entire evolutionary trajectory.

In reality, however, most mathematical models focus on evolutionary dynamics in just one layer. Genetic models aim to directly calculate the change in genetic frequencies from G_i to G_{i+1}, while phenotypic models are concerned exclusively with the shift in phenotypic features from P_i to P_{i+1} (Fig. 1.2, dashed arrow). Such calculations are achieved by building a proper *transition function*. Let X denote either phenotypic or genotypic profile of a population, and $\Delta X := X_{i+1} - X_i$ the change of the population profile between two generations. A transition function has the form

$$\Delta X = f(X; \ \alpha, \beta, \dots)$$

where $\alpha, \beta, \dots$ are *parameters* of the function that summarize the developmental or evolutionary factors at work in the steps $T_1 \sim T_4$ above. If the function and parameters well capture these processes, one can successfully derive the evolutionary change based on the present state X of the population. The task of population genetics thus boils down to identifying the form of the transition function and determining its parameters for the evolutionary process under study.

As a concrete example, consider the following one-locus population genetics model that describes the change in the population frequencies p,q of alleles A,a in response to selection:[1]

$$\Delta p = f(p,q; \mathbf{w})$$

$$= \frac{pq[p(w_{AA} - w_{Aa}) + q(w_{Aa} - w_{aa})]}{p^2 w_{AA} + 2pq w_{Aa} + q^2 w_{aa}} \tag{1.1}$$

(Note that the lowercase p here denotes a *genetic* frequency and not a phenotypic one as denoted by the capital P above.) Here, the frequency change Δp is determined from the current allele frequencies (p and q) and three *fitness parameters* $\mathbf{w} = (w_{AA}, w_{Aa}, w_{aa})$, which represent the chance of survival from

---

[1] This Element considers only infinite-population models, where population dynamics is deterministic with no drift.

birth to the adult stage; so if eight out of ten AA individuals survive to reproduce, $w_{AA} = 0.8$. Since the survival rate reflects the strength of selection, Eqn. 1.1 can be thought of as describing how the population frequencies change – that is, evolve – in response to selection. Just to give an idea, suppose further $w_{Aa} = w_{aa} = 0.5$, that is, half of Aa and aa individuals die before reproduction. We also assume the initial population contains the same amount of A and a alleles, so that $p = q = 0.5$. When we plug these figures into Eqn. 1.1, the change in the frequency is calculated as follows:

$$\Delta p = \frac{(0.25)[(0.5)(0.8 - 0.5) + (0.5)(0.5 - 0.5)]}{(0.25)(0.8) + 2(0.25)(0.5) + (0.25)(0.5)} \approx 0.065 \qquad (1.2)$$

Thus, the frequency of A allele will increase to about 56.5%, in response to selection favorable for AA individuals. This process can be reiterated to yield the population frequencies of arbitrary generation (Fig. 1.3).

Although this model only deals with selection, other evolutionary factors such as mutation, migration, randomness (drift), population structure, and so

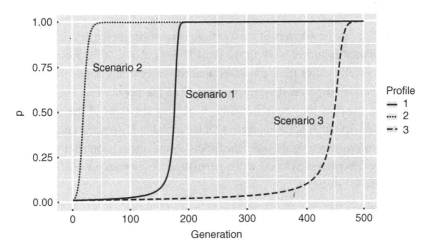

Figure 1.3 Simulation of evolutionary trajectories generated from repeated applications of Eqn. 1.1 with three different fitness parameters, all starting from the initial frequency $p = 0.01$. **Scenario 1** ($w_{AA} : w_{Aa} : w_{aa} = 0.8 : 0.5 : 0.5$) is the example in the main text, and in this case A almost reaches fixation in less than 200 generations. A evolves even faster in **Scenario 2** ($w_{AA} : w_{Aa} : w_{aa} = 0.8 : 0.65 : 0.5$) where the heterozygote fitness is intermediate (i.e., no dominance). The adaptive evolution slows down in **Scenario 3** ($w_{AA} : w_{Aa} : w_{aa} = 0.8 : 0.65 : 0.65$) where the fitness difference is less significant.

forth can be incorporated into the model, and their relative importance in evolutionary processes can be assessed by comparing the model's prediction and actual observations. If, for example, a target population did not respond as predicted by Eqn. 1.2, we may infer either that our fitness estimate was incorrect (i.e., wrong parameters) or that other evolutionary forces not included in the model were at work (wrong functional form).

Models can also be used the other way around to estimate parameters. From the early nineteenth to the mid-twentieth centuries in a forest near Manchester, a dark (melanic) form of peppered moth, *Biston betularia*, increased its frequency at the expense of the original light-colored form. When a model like the one above was fitted to the actual records of frequency change, it was estimated that the light-colored moths had two-thirds the survival rate of melanic moths, so that $w_{light} : w_{dark} = 2 : 3$. In these ways, population genetics models have enabled prediction, estimation, and testing of evolutionary dynamics and factors in a quantitative, hypothetico-deductive fashion.

These rigorous and formal treatments of evolutionary dynamics took on great significance not only for their predictive value, but also for their metascientific implications for the status of evolutionary theory. Unsurprisingly, the rise of modern evolutionary theory has generated much philosophical reflection on its theoretical status, especially its integrity and relationship to the physical sciences (Smocovitis, 1996, ch. 5). One of the primary contributors to the Modern Synthesis, J. B. S. Haldane (1931, p. 150) stressed that "biology must be regarded as an independent science with its own guiding logical ideas, which are not those of physics." What concerned him was the reductionist atmosphere of the time engendered by adamant physicists like Lord Kelvin, who infamously proclaimed that every natural phenomenon eventually could be explained by combinations of physical laws, making the rest of the sciences just applied physics or even "stamp collecting." In order for evolutionary biology to be an autonomous and respectable science, thought Haldane, it must have its own set of laws or "guiding logical ideas" that are as rigorous as those of physics but not reducible to them. Population genetics, with its quantitative treatment of evolutionary change, was expected to provide just such laws of evolution. It is for this reason that Fisher (1930) called his formula on the rate of adaptive change the "fundamental theorem of natural selection" and likened it to the second law of thermodynamics. This basic formula, Fisher proclaimed, holds true "of any organism at any time," and the existence of such universal laws of evolution was taken to establish evolutionary biology as a rigorous scientific discipline with its own principles.

The search for universal laws has led to abstraction and distillation of the logical essence of the evolutionary process. Darwin formulated evolution by natural selection as a necessary consequence of three conditions, namely (i) phenotypic variation (organisms in a population are not all the same but differ from each other), (ii) associated fitness variation (difference in phenotype results in a difference in organisms' chance of survival and reproduction), and (iii) heritability (offspring resemble their parents) (Darwin 2003, ch. 4; Lewontin 1970). Presented as such, the argument does not make any substantive assumption about the biology of an evolving population. Indeed, any collection of entities – even inanimate things – that make their "copies" at differential rates evolves through the process of natural selection according to this construal. For this reason some biologists have concluded that adaptive evolution in its purest form is a logical fact, holding true of any population in arbitrary environmental circumstances as long as it satisfies very general premises (e.g., Endler, 1986; Ridley, 2004)

The Darwinian syllogism derives adaptive evolution from the three conditions. But can we be more precise and calculate *how much* a particular population quantity, say its mean height or weight, changes due to selection and other evolutionary factors? The answer is yes. The so-called *Robertson-Price identity* or simply the *Price equation* (Robertson, 1966; Price, 1970) gives the change in the mean of any phenotypic character between two generations, expressed by $\Delta \overline{Z}$, as a statistical function:

$$\Delta \overline{Z} = \mathrm{Cov}(W, Z')/\overline{W} + \overline{\delta Z}. \tag{1.3}$$

The variables W, Z, Z' and δZ in the equation quantify properties of individuals in the population. The fitness W is simply the number of offspring, so if an individual has two offspring its W value is 2.[2] Z quantifies any phenotypic characteristic of an individual, while Z' is the average phenotypic value of its offspring; so if we are interested in weight, and the above individual weighed 9 grams while its two offspring weighed 8 and 12 grams, its values of Z and Z' are 9 and 10 respectively. Finally, δZ is defined as $Z' - Z$, that is, the difference between the phenotypic value of a parent and the average phenotypic value of its offspring, and is often interpreted as the *transmission bias*. For the above individual, this value is $10 - 9 = 1$. That is, the offspring produced from this individual were on average 1 gram heavier.

With this in mind, the Price equation can be explicated in two parts. The first term $\mathrm{Cov}(W, Z')/\overline{W}$ is the *covariance* of the fitness and the average offspring phenotype divided by the *mean* fitness. Since the mean fitness is never negative, the sign of this term is entirely determined by how the two variables

---

[2] Here we assume asexual reproduction for expository convenience. In the case of sexual reproduction fitness must be divided by two to take into account the fact that each offspring has two parents.

W and $\acute{Z}$ covary. If the covariance is positive and (as in the above case) individuals having heavier offspring tend to leave more offspring, this term tends to push up the mean population weight, but the reverse if those individuals tend to procreate less. This term thus captures Darwin's basic idea that a characteristic contributing to fitness (that is, number of offspring) will spread in a population. The second term $\overline{\delta Z}$, in contrast, is the mean transmission bias and measures whether and to what extent offspring on average differ from their parents, irrespective of selection (because this term does not contain fitness W as a factor). Combining these terms together, the Price equation calculates the mean phenotypic change $\Delta \overline{Z}$ of a population from the quantitative expressions of selection and the transmission bias.

The striking fact about Eqn. 1.3 is that it is obtained through pure deduction as a mathematical theorem that follows from the basic axioms of probability theory and the definitions of mean and covariance (see, e.g., Okasha, 2006; Frank, 2012, for accessible expositions). Free from any biological or empirical assumption, the equation is thus applicable to any population change, just as $7 + 5 = 12$ holds true of any countable objects. Due to this logical austerity and universality, the Price equation has played essential roles in theoretical biology (Frank, 1995, 2012; Luque, 2017) and is often touted as "the most fundamental theorem of evolution" (Queller, 2017).

All the refinement and purification of evolutionary principles, however, has invoked a philosophical puzzle: why can such mathematical theorems tell anything at all about actual and concrete evolutionary processes? In part, this is an echo of an old philosophical conundrum dubbed by Eugene Wigner (1960) as "the unreasonable effectiveness of mathematics in the natural sciences." Ever since Galileo, scientists have made use of mathematics to study empirical and causal structures of the world, evidently with great success. But why do mathematical theories, which are seemingly constructed "in our head," describe the world, which is obviously "outside our head"? This question has been asked time and again by, to name just a few, Descartes, Kant, and the logical positivists, each in response to the contemporary developments of the natural sciences: Galilean physics (in the case of Descartes), Newtonian mechanics (Kant), relativity theory (logical positivists), and quantum mechanics (Wigner). Just as the successes of these physical theories have invited metaphysical reflections on the conditions that would sanction the use of mathematics in the physical sciences, the development of population genetics in the twentieth century naturally led philosophers to a similar inquiry regarding the role and nature of mathematical reasoning in evolutionary studies.

In effect, the unreasonableness of the effectiveness of mathematics is even more acute in the case of evolutionary biology due to the aforementioned a priori character of its fundamental principles. Physics does extensively use abstract mathematical formulae, but the ultimate arbiter is Nature. The truth or falsity of Newton's or Einstein's laws is not guaranteed by mathematics alone. You need observations and experiments to decide which, if any, are true – otherwise, Eddington could have better sat in his armchair and calculated, rather than mounting his famous expedition to the island of Principe. In this sense, these equations are not themselves products of mathematics but rather mathematical expressions of empirical hypotheses. In contrast, we have just seen above that the fundamental principles of evolution are often considered to be mathematical theorems that hold without any empirical assumptions. We don't need any observation or experiment to bear out the Price equation, because its truth is entailed by probability theory alone. But if so, it is all the more puzzling why such a priori statements could sustain hypotheses about historical origins of species or predictions about future evolutionary trajectories. This puzzlement gave rise to a suspicion that evolutionary theory is in fact not an empirical theory with falsifiable hypotheses but rather an elaborated set of tautologies (Smart, 1959; Popper, 1974). The apparent a priori-ness of evolutionary principles also casts a shadow on Haldane's hope for the autonomy of biology, for if the putative laws of evolution turn out to be mathematical facts that would obtain regardless of any empirical conditions, it would be utterly unclear why they could serve as the basis for the autonomy and integrity of *biological* sciences.

At stake here is not just the empirical nature but also the predictive capability of evolutionary theory. Darwin's principle of natural selection arrives at adaptive change from the premise that individuals in a population differ in their capacity to survive and reproduce and that the capacity is heritable. This reasoning is apparently *ampliative*, that is, its conclusion seemingly delivers new information that was not included in the premises. In other words, evolutionary change is *predicted* from the heritable differences in fitness, and, as we have seen, this predictive ability of Darwin's principle, backed up with quantitative formulations of population genetics, played a central role in its acceptance among biologists in the early twentieth century. But how is such ampliative reasoning possible if the underlying principle was a logical or mathematical truth? Logical deductions may explicate the information contained in the premises, but never extend our knowledge beyond them. Hence, should Darwin's principle be a kind of logical deduction, it would never be able to predict an adaptive change before it actually happens – what it could do would be, at most, relating a past change to the selective and hereditary conditions. The deductive outlook of evolutionary principles thus casts a serious doubt on the ampliative nature and predictive ability of evolutionary theory.

The mathematization of evolutionary principles, therefore, presents both an opportunity and a challenge for evolutionary theorists. On the positive side, it saved Darwin's theory of natural selection from its early criticisms and provided the theory with quantitative formulations to study evolutionary change in a rigorous and precise fashion. On the other hand, the abstract and logical nature of these principles has come to cast a shadow on the empirical status and predictive ability of evolutionary theory. For these opposing reasons the use of mathematical models in population genetics has attracted the attention of biologists, philosophers, and skeptics who are in one way or another interested in the nature of evolutionary theory. To the eyes of skeptics the issue signaled that Darwin's theory is not actually a respectable scientific hypothesis but rather a "tautology" (see the next section); the pioneering figures of the philosophy of biology, in contrast, recognized it as the unique feature of evolutionary theory that cannot be properly evaluated by the standards of other physical sciences. Either way, giving a proper place to the role of mathematics was considered to be key both to understanding the structure of evolutionary theory and to establishing it as an empirical, predictive, and autonomous scientific theory that merits its own philosophical attention.

This Element introduces the reader to this traditional issue in the philosophy of biology, with a particular focus on its implications for the metascientific understanding of evolutionary theory. The first part (Sections 2 and 3) presents two major views of the mathematical nature of evolutionary theory that have been developed in the past few decades. Section 2 kicks off the discussion from the 1980s, when the problem elicited classic works that came to define the "received view" in the philosophy of biology (e.g., Mills and Beatty, 1979; Brandon, 1981; Sober, 1984; Rosenberg, 1985). According to this view, mathematical models of evolution acquire an empirical purport through proper *interpretations* of the parameters and concepts therein. Of particular focus among these concepts was *fitness*, and we will see how its interpretations, called the propensity interpretation and the supervenience thesis, were thought to make mathematical formulae of evolution empirical and ampliative. The received view, however, came to be challenged around the turn of the century by an alternative called *statisticalism*, which claims that population genetics is best understood as a branch of mathematical statistics (Matthen and Ariew, 2002; Walsh et al., 2002; Pigliucci and Kaplan, 2006; Walsh et al., 2017). In this construal, evolutionary models apply to an actual population just as statistical models do to sampled data. Section 3 expounds this view and examines its implications for the explanatory and inductive structure of evolutionary theory.

After a survey of the existing literature, the second part sketches the author's own take on the issue. The previous views, either the received view or statisticalism, have presumed that mathematical models of evolution are purely a priori theorems distinct from empirical statements. The sharp distinction between mathematical and empirical statements, however, has been questioned ever since Quine's (1951) criticism of the "two dogmas of empiricism." Drawing on Quine and another critic of logical empiricism, Patrick Suppes, Section 4 offers a holistic picture of evolutionary theory in which the mathematical models of population genetics and empirical observations in physiology or ecology form a contiguous network to explain and predict evolutionary phenomena. The key component in this network is the *causal graph theory*, a formal framework that bridges probability and causality. Section 5 fleshes out this idea and shows how causal models of evolving populations connect the mathematical and biological assumptions of evolutionary theory and also serve as the *uniformity of nature* that justifies inductive reasoning and predictions of evolutionary changes. Finally, Section 6 concludes with a brief review of the roles of formalization for studies of evolution and sketches its implications for future philosophical discussions on the structure, methodology, and ontology of evolutionary theory.

2 The Received View

Although the history of the metascientific reflection on evolutionary theory is as old as Darwin's theory itself, it did not become a major topic in the professional philosophy community until the latter half of the twentieth century, when major textbooks and monographs appeared to define the field as a branch of the philosophy of science.[3] The belated uptake is often attributed to the historical background of the philosophy of science, which from its conception at the Vienna Circle has been dominated by reflections on theoretical physics. The disparagement of nonphysical science is well exemplified by J. J. C. Smart's (1959) controversial paper "Can Biology Be an Exact Science?" where his answer was a categorical "no." For Smart and other contemporary philosophers, the hallmark of an exact science is the existence of its own laws, defined as generalizations that are both universal (i.e., have no exception) and empirical (discovered by a posteriori investigations). Einstein's field equations and the Schrödinger equation are paradigmatic examples of laws that epitomize the theoretical rigor of, respectively, the general theory of relativity and quantum mechanics. For Smart, biology lacks such universal laws because all nontrivial generalizations about organisms are either liable to exceptions or reduced to tautological statements such as "survival of the fittest." He thus concludes that biology is all about natural history or applications of more fundamental laws of

---

[3] For instance, Ruse (1973); Hull (1974); Sober (1984); Rosenberg (1985).

chemistry or physics, and in this regard it is more like radio engineering rather than pure sciences. An implicit corollary of this is that there can be no philosophy of biology just as there is no philosophy of radio engineering. It is no wonder that Smart's skepticism presented a significant challenge to pioneering figures of the philosophy of biology. To prove the raison d'être of their own field, philosophers of biology had to show that the subject of their philosophical analysis was not only an empirically testable and scientifically respectable theory but also an autonomous discipline irreducible to more basic sciences.

Smart's challenge obviously points to the conundrum we saw in the previous section, namely that the fundamental principles of evolution can be formulated in terms of a logical syllogism or a "mathematical tautology" like the Price equation. In the philosophical literature the issue has been known as the *tautology problem*, named after the notorious charge that the Darwinian principle of natural selection, summarized as "survival of the fittest," is tautological if "fitness" is just another name for better survival and reproduction rates. This framing suggests that the issue is a semantic problem. A sentence is judged tautological or not according to the meaning of the terms therein: for example, "all bachelors are unmarried" is a tautology because the word "bachelor" means, by definition, an unmarried man. If so, the allegation of tautology against the above slogan may be dispelled by providing a proper definition of the concepts therein, in particular that of fitness. This prospect led philosophers to adopt *interpretation* as a general strategy to tackle the conceptual problems. That is, determining the correct meaning of central concepts of evolutionary theory, most notably fitness, was believed to be key to turning its mathematical or logical rules into empirical laws. The philosophical investigation in this line came to fruition in two major accounts of fitness, the propensity view and supervenience thesis, which address the problems of tautology and reductionism respectively. This section spells out these two interpretations in detail and examines how they address Smart's challenge.

2.1 The Propensity Interpretation

How did philosophers of biology respond to the charge that evolutionary theory is based on a priori theorems or "tautologies"? In fact most of them bit the bullet, conceding that the essence of evolutionary theory can be summarized by a purely deductive formula. Robert Brandon (1981, p. 433) admitted that the principle of natural selection "has no empirical content of its own ... It is simply an application of probability theory to a biological problem." This was echoed by Elliott Sober (1997, p. 458): "the process of evolution is governed by models that can be known to be true a priori ... it turns out to be a mathematical truth." Alex Rosenberg (1985, p. 128) went so far as to claim that "Evolutionary theory cannot be anything like a theory in physical science . . . Like Euclidean geometry, it can be treated as a system of pure definitions or 'a body of tautologies'."

The philosophers, however, denied that the a priori nature of evolutionary principles makes the *entire* evolutionary theory tautological; for although the derivation of these principles may not require any empirical assumption, their application does. To predict which type of organisms will spread in the future with some mathematical model, one has to know which phenotypic character gives its bearer survival and/or reproductive advantage, and this is not something one can know a priori. Measurements or experiments are required, and these empirical procedures make an abstract evolutionary model applicable and relevant to a concrete biological population. Philosophers thus have adopted a divide-and-conquer strategy to cope with the charge of tautology, dividing evolutionary theory into theoretical deductions and empirical applications and grounding the empirical nature of the theory on the latter components: "the empirical biological content of Darwinian evolutionary accounts lies in these instantiations or applications" (Brandon and Beatty, 1984, p. 343).

This strategy is nicely illustrated by Sober's (1984) distinction between consequence and source laws. The term *consequence laws* is Sober's parlance for mathematical formulae of evolution that calculate population change from certain initial conditions and parameters. The population genetics model we saw in Section 1 (Eqn. 1.1), which gives the change in gene frequency as a function of initial gene frequencies (p,q) and genotypic fitness (w_{AA}, w_{Aa}, w_{aa}), is a typical example. As the name suggests, consequence laws focus on deriving evolutionary consequences from a given setup, but they are silent about where their initial conditions and parameters come from in the first place. In order to apply these mathematical models to study the evolution of an actual population, one needs to estimate the parameters via another set of laws called *source laws*. Source laws assess ecological, demographic, life-history, and other factors and convert them to numerical values that can be plugged into a given consequence law. This is an empirical procedure par excellence, which involves careful observations of, or laborious experiments on the target population and its environment. Since evolutionary explanations are combinations of these two kinds of laws, then even if consequence laws are a priori, "[t]he explanation *as a whole* is empirical because *other* components of it are" (Sober, 1984, p. 79; italic original).

Source laws thus breathe empirical life into cold mathematical formulae by supplying their parameters and variables with empirical content based on observations and measurements made on the target population. But how is this achieved? Measurement maps concrete properties of organisms or environments to abstract quantities. Hence, in order to determine some quantity in a mathematical model, we must first identify the biological property to which it corresponds; that is, we need to know what that quantity or concept actually *means*. In order to use Newton's formula $F = ma$ to study the motion of a particular body we need to know what the force, mass, and acceleration of that body are; otherwise, we will be at a loss as to where to look to determine their values. Likewise, philosophers have thought that the application of evolutionary formulae presupposes a proper understanding of the concepts used therein; that is, we have to correctly *interpret* these concepts in the first place. Of particular concern is the concept of fitness, which arguably plays the central explanatory role in the theory of natural selection. Mathematical models of adaptive evolution calculate population change as a function of fitness, such that organisms with higher fitness survive and leave more offspring in the next generation. This process is often summarized as "the survival of the fittest" – but who are the fittest? Should the fitness of an organism be defined by the *actual count* of its offspring, the above statement collapses into a tautology – those who procreate well procreate. Hence if evolutionary theory is to say anything more substantive than just A is A, fitness must be interpreted differently, that is, it must mean something other than actual number of offspring.[4]

But then what? The philosophers' answer was that fitness denotes not an actual outcome, but the *propensity* of an organism to survive and reproduce (Mills and Beatty, 1979; Brandon, 1981; Sober, 1984). A propensity is a property of a thing that disposes it toward a certain behavior or outcome under specified conditions. Fragility, for example, is the propensity of an object to easily shatter when hit by hard material. The fairness or biasedness of a die is its propensity to roll towards certain outcomes. As with inanimate things, we can conceive of organisms having various propensities in different degrees: some organisms may be more "fragile," or frail, than others so that they are less likely to survive severe environmental conditions such as drought, or their reproductive outcome may be "biased" so that they tend to have more (or fewer) offspring than average. There are many other such propensities that could affect an organism's chance of survival and reproduction, and fitness can be conceived of as a general property that summarizes them all; that is, it is

---

[4] For some philosophers, the fate of the Darwinian theory as an empirical science hinges just on that – that is, whether one can provide a correct interpretation of fitness or not. Rosenberg (1985, p. 129) thus claims "Biologists must face the problem of providing a noncircular account of the meaning of the key terms of evolutionary theory or forswear its explanatory power altogether . . . If we cannot provide respectable accounts of the meaning of key terms of the theory that are independent of it, then we will have to surrender all hope of showing that the Darwinian theory is in fact a scientific theory." A similar view can be found in Mills and Beatty (1979) and Brandon and Beatty (1984).

defined as an organism's overall propensity to survive and reproduce in a certain way under a given environment.

What underlies the notion of propensity is the distinction between actuality and potentiality. A wine glass that has never shattered can still be fragile, just as flipping a fair coin ten times may yield no heads, for what determines something's fragility or fairness is not outcomes but rather its potentiality or structure. In the same vein, there is no contradiction if organisms with higher fitness end up faring worse than those with lower fitness. Such a discrepancy between a propensity and an actual outcome may result from bad luck. But if that is the case, it would not be tautological to say that these same organisms, when they are actually successful, have survived or reproduced better *because* they have a propensity to do so, that is, have higher fitness. Better yet, the statement seems to be explanatory in the sense that it provides a reason as to why they, and not their rivals, won the struggle for survival. Such an explanation would be no less legitimate than resorting to fragility or fairness to explain an eventual shattering of the glass or getting five heads when flipping ten fair coins. If fitness *qua* propensity is nothing but the "biasedness" of an organism to survive and leave offspring, then it should provide an explanation of the actual success or failure of that organism just as an unfair coin explains a lost bet. And explanation by natural selection is just like that: it explains *actual* frequency changes of organisms on the basis of their *potential* to survive and reproduce. Note that such an explanation would be void should fitness be defined by the actual change, the very target it purports to explain. Hence the propensity interpretation, with its connotation of potentiality, endows a real explanatory value to evolutionary theory, or so it is argued.

Another feature of propensity is that it comes in degree. A wine glass, say, is more fragile than a coffee mug. It is also stochastic. When I say that the glass is fragile, I do not mean that it never fails to shatter whenever hit by hard material, but just that it shatters with high probability. Likewise, although fitter organisms tend to survive or reproduce better than less-fit ones, whether they really do so or not is a matter of chance. How do we represent such differential degrees of stochastic properties? The philosophers' answer is that propensity fitness denotes *statistical expectation* of survival or reproduction (Mills and Beatty, 1979; Sober, 1984). Let us see this with survival selection first. Whether an organism survives to its sexual maturity or not can be expressed by a binary variable Y, where $Y = 1$ means survival and $Y = 0$ death. Since selection is a stochastic process, we can think that each outcome occurs at probability $P(Y = 1)$ and $P(Y = 0)$ respectively, which jointly define the *probability distribution* $P(Y)$ over variable Y. Formulated as such, survival selection is just like coin flipping, where organism or genotype correspond to coin or type of coin with different biases. What is biasedness? Statistically speaking, it is nothing but the frequency with which a coin is expected to land on heads (or tails). Likewise, for each organism or genotype one can define its *survival fitness* as the expected value of Y:

$$E(Y) = \sum_{y=0}^{1} P(Y = y) \cdot y$$

which in this case reduces to $P(Y = 1)$, that is, the probability of survival. Fitness in reproductive selection can be treated in the same fashion. Let W be another variable that denotes the number of offspring, so that $W = 0$ if an organism has no offspring, 1 if one, and so on. Again, how many offspring an individual can have is a stochastic process, determined by probability distribution $P(W)$. Its expected value is

$$E(W) = \sum_{w=0}^{\infty} P(W = w) \cdot w$$

which defines *reproductive fitness*.[5]

Fitness, therefore, is a propensity of an organism formally expressed by its expected survival rate or number of offspring. For instance, the genotypic fitness w_{AA} in the population genetics model (Eqn. 1.1) in the previous section is nothing but the expected survival $E(Y)$ of AA individuals, or more formally the conditional expectation $E(Y|AA)$, and likewise for the other genotypic fitnesses w_{Aa} and w_{aa}. Defined as such, the model is no longer tautological, for the actual survival rate of any given genotype may well differ from its expectation, in which case the calculation from the model does not exactly match the actual evolutionary response. But overall, fitter genotypes are expected to survive and reproduce better than less-fit ones, and thus will spread within a population in the long run – just as we will eventually see more of one side than the other when we keep flipping a biased coin. And if the biasedness of the coin predicts and explains its long-term frequency, so does the fitness *qua* propensity/expectation explain changes in population frequencies. In this way, the propensity interpretation renders mathematical models of natural selection "not only empirical, but explanatory," as its proponents argue (Brandon and Beatty, 1984, p. 343).

---

[5] The probability distribution $P(W)$ here is called the *fitness distribution*. Although philosophers have almost exclusively focused on its expectation, nothing in the propensity interpretation precludes higher moments such as variance or skewness from the consideration of fitness (e.g., Pence and Ramsey, 2013). The propensity interpretation thus understood generally is the claim that the concept of fitness concerns parameters, and not sample moments, of the fitness distribution.

Above I noted that in order to measure some quantity one must first determine its meaning. The propensity view is primarily a semantic thesis that responds to this question regarding the meaning of fitness parameters (to the effect that they mean not the actual but expected survival rate or offspring number). But it also has epistemological implications for estimating the fitness, giving rise to two distinct strategies for building source laws.

The first approach to estimate fitness, which is immediately suggested from the above discussion, is statistical estimation (Mills and Beatty, 1979). The expected values are what statisticians call *parameters*, which characterize a chancy setup but always stay behind the scenes. They are like configurations within a machine that operates stochastically – which are never directly observed but can only be indirectly inferred from observed data. The art of inferential statistics consists in giving estimates of latent parameters that characterize a probability model from observed *sample statistics*, within a certain error bound. For instance, the biasedness of a coin may be estimated by tossing it a number of times. If we toss the coin 1,000 times and get 524 heads, we may infer that the coin is not so terribly biased – that is, its true probability of getting heads does not deviate largely from 0.5. Likewise, we may estimate the reproductive fitness of one type of organism from its average count of offspring. If, upon an examination of 100 organisms each, *AA* individuals have 5.7 and *aa* individuals 1.8 offspring on average, we may infer that *AA* has a higher (expected) fitness than *aa*. The inference is fallible: it might have just so happened that our sample contained a disproportional number of lucky *AA* individuals and unlucky *aa* ones, or that the obtained difference is not just statistically significant. Hence there is always a chance of error, but statistics gives us ways to handle such uncertainty and to assess the reliability of given statistical estimates.

Statistical estimation does not presuppose any knowledge about the physical nature of the object. All we need to conclude the biasedness of a coin through repeated tosses is just head counts and none of its physical properties. But we can sometimes detect a possible bias through a physical inspection of a coin, say by finding it bent, without actually tossing it at all. The second approach to assess fitness exemplifies this line of reasoning and proposes to estimate fitness on a physical basis. Above I suggested that something's propensities are based on its structure: glass, for example, is fragile due to its material nature. Likewise, if organisms differ in their physical or physiological conditions–for example, if some can run faster or have better metabolic rates than others–then we may infer that the former have a higher fitness, that is, they are better disposed to survive and reproduce. In general, an organism's design provides valuable information about its performance, just as the shape and surface of a coin give a clue as to its biasedness. Ornithologists conjecture that colorful male song birds attract more females than less-colorful ones, and functional morphologists evaluate various forms of fins in their capacity to produce underwater propulsion. By so doing, they estimate the fitness based on the morphological properties of an organism. Such *design analysis*, therefore, should count as the second type of source laws to estimate the propensity fitness (Gould, 1976; Beatty, 1980; Brandon and Beatty, 1984; Sober, 1984).

In sum, the so-called tautology problem has two aspects. Its semantic aspect concerns the analyticity of the concept of fitness and is typified by the claim that the Darwinian principle of "survival of the fittest" is a mere tautology. The propensity interpretation responded to this charge by redefining fitness as a propensity represented by the expected rather than actual number of offspring. Since fitness thus understood as a potentiality need not equal the actual success, "survival of the fittest" is no longer analytic. The other aspect is epistemological and concerns the a priori-ness of evolutionary equations that are allegedly proven as mathematical truths. This charge was addressed by focusing on the process of applying these equations, or more specifically on the role of the source laws that provide empirical estimates of their parameters. Of particular interest among these parameters is fitness, and we have seen statistical inference and design analysis as two major methods to estimate the fitness parameter/ expectation.

Although the name "tautology problem" suggests that the issue just concerns the logical status of evolutionary theory, what is really at stake is its inductive nature – how can we infer and predict future evolutionary trajectories with apparently a priori mathematical principles? As illustrated in Fig. 1.3, a population genetics model concludes a drastic change in population frequencies from repetitions of relatively weak selection. But what would guarantee such an extrapolation of one evolutionary model over hundreds or thousands of generations? First, the recursive application assumes the same model holds true during the scope of the inference, which in turn requires that its parameters, including most notably fitness, capture stable and inherent properties of the target population rather than transient statistics peculiar to each generation. Second, these parameters must be reliably estimated from limited samples obtained from one or a few generation(s); in Humean terms, the two conditions for an evolutionary model to be inductively successful are that the model and its parameter reflect the uniformity of nature and that this uniformity is inferred from a few observations. The propensity interpretation and source laws answer to these two demands. Grounded on the physical structure of an object, a propensity serves as a *projectible* property (Goodman, 1955) that warrants inductive reasoning. A biased coin stays biased by itself, so that we can reliably infer the coin keeps landing disproportionally. Likewise, as long as fitness is inter-preted as a propensity, one can "project" the same model to predict a future evolutionary trajectory. The source laws, on the other hand, provide empirical means to determine such a projectible property through statistical estimation or design analysis. Both of them are needed to infer evolutionary dynamics with mathematical models, and in this sense the propensity interpretation and the source laws serve as the two legs on which evolutionary theory stands as an empirical and inductive science.

2.2 Supervenience Thesis

While the propensity interpretation was targeted at the tautology problem, the supervenience thesis addresses the other problem of *reductionism*, the view that evolutionary theory or more generally biological sciences are in principle reducible to more basic sciences such as chemistry or physics. Against this challenge, philosophers of biology defended the epistemic value of evolution-ary theory by claiming that its key explanans, fitness, supervenes on multiple physical structures and reveals general patterns that are not visible in the reductionist picture.

In the philosophical literature, the concept of *supervenience* denotes one-to-many ontological dependence relationships. A prime example is the relation-ship between temperature and underlying molecular states. Temperature *is* molecular motion, so they are ontologically the same but not on a par: tempera-ture *depends* on the underlying molecular states in the sense that a difference in the former entails that of the latter but not vice versa, for distinct molecular states may correspond to one and the same temperature. Supervenience captures this one-to-many relationship. In general, a property A supervenes on another property B only if a difference in A entails that of B, but not necessarily vice versa.[6]

Propensity is a supervenient property *par excellence*. Fragility, for instance, is realized by various physical structures or compositions such as glass, cera-mic, or carbon, so that my coffee mug and the window of my office may be equally fragile despite their material difference. Likewise, fitness qua propen-sity supervenes on multiple phenotypic and genotypic structures; that is, the same fitness value may be assigned to organisms with different behavior, morphology, or genes. A southern African lion and a green sea turtle have

---

[6] Sometimes supervenience is distinguished from multiple realizability, the latter being defined as a one-way supervenience: A is multiply realized by B if A supervenes on B but the converse does not hold (B does not supervene on A). In the philosophical discussion of fitness, however, supervenience is almost exclusively used in the sense of multiple realizability, so I do not distinguish them here.

very different forms, habitats, and life histories, but that does not prevent them, logically at least, from having the same expected number of offspring that survive to sexual maturity – that is, the same fitness value. So difference in phenotype or environment does not imply difference in fitness, but the converse does not hold: if the fitnesses of two individuals differ systematically, we infer that there must be some phenotypic or environmental factor responsible for their differential success (note that we are dealing with fitness as expectation, so contingent factors are excluded). Hence the relationship between fitness and phenotype-cum-environment is one-to-many, or in other words, the former supervenes on the latter.

The concept of supervenience has attracted philosophers with an antireductionist slant, for it seems to attest that the explanatory value of some macro properties is not exhausted by micro properties. Propensities are paradigmatic examples of such macro properties. As we have seen, propensities do explain the behaviors of a physical object; but reductionists may argue that its microscopic structure, described in terms of chemistry or particle physics, offers a much deeper and more precise explanation of the same phenomenon. Then why should we bother with propensities? The answer is that supervenient propensities provide general categories with which we can *uniformly* explain or predict the behavior or outcomes of objects of heterogeneous nature. For instance, by labeling them fragile we can explain characteristics of various materials with different chemical structures. Such general patterns or commonalities are usually not obtainable from close-up, microscopic studies of individual materials, and herein lies the epistemic value of supervenient properties (Fodor, 1974; Putnam, 1975). In the same vein, the supervenient nature of fitness suggests the way evolutionary models get generalized (Sober, 1984, 1993; Rosenberg, 1985). If fitness supervenes on different phenotypic and environmental conditions, and the fitness parameter, among others, determines evolutionary dynamics, it opens the possibility of applying the same evolutionary model to different populations. Whether it is a population of fruit flies or zebras, the theory of natural selection predicts or explains that fitter organisms survive and spread based on the same principle. Hence the supervenience of fitness on physical properties allows evolution of various populations under different conditions to be *unified* under one or a few principles. On the other hand, reductive explanations, detailed and precise though they may be, lack such unifying power. A Laplacian demon who has complete knowledge of the world and infinite computational power may well predict any future state of any particular population by referring only to physical laws and conditions, but he has no idea about *common patterns* exhibited by different evolving populations. Evolutionary theory deals with such broad patterns that emerge only at a macro level through the supervenience of fitness, and in this regard it has its own explanatory values – generality and unification – that are not obtainable from more fundamental sciences such as chemistry or physics.

The supervenience thesis is also favorable from the viewpoint of the ontological orthodoxy of evolutionary theory, known as *population thinking*. According to Ernst Mayr, one of the most influential evolutionists in the twentieth century, Darwin revolutionized not only the science but also the general ontology of living things: that is, he changed our understanding of not just how organisms and species arise, but also what they are. The traditional, pre-Darwinian biology presumed that organisms or species embody distinct types or essences, which supposedly determine their nature and behavior. This assumption of a universal and stable nature, however, was discredited at the advent of the Darwinian theory of evolution, which requires any population and species to be inherently variable and malleable for evolution by natural selection to be possible (Mayr, 1975). Everlasting evolution means that everything in biology is variable and transient, so there can be no fixed biological kinds.

The ontological shift and the denial of biological kinds, however, cast a shadow of doubt on the notion of laws in evolutionary theory. We have so far discussed the roles of laws – source laws and consequence laws – in studies of evolution. But what are laws in the first place? The traditional concept of lawhood is tightly linked to that of *natural kinds*. Laws are supposed to relate or describe natural kinds – for example, laws of chemistry determine the properties of chemical kinds such as iron, which combines with oxygen, reacts with acid to produce a salt and hydrogen, and so on. But if population thinking is right and there is no such thing as biological kinds, then what do the source and consequence laws take as their relata? Does that mean they are not laws after all, and thus evolutionary theory lacks its own inductive principle?[7] The supervenience thesis seems to offer an ideal solution to this problem. Fitness, as a supervenient property, does not designate the intrinsic nature of an organism, but rather its relationship to other organisms and its environment. To say that some fruit fly has a fitness value *w* does not presuppose an essence, otherwise it would be impossible for them to have different fitness values in a different environment or generation. The fitness of an organism is determined not just by its own physical characteristics but also by its environment and other individuals in the population, and in this sense it is a relational or population property. The supervenience thesis implies that evolutionary laws take such population properties

---

[7] Indeed, this was the conclusion that Mayr (1982), among other philosophers like Beatty (1995), drew from population thinking. They denied the existence as well as the explanatory role of laws in evolutionary biology and claimed that evolutionary theory is fundamentally a historical science distinct from other nomological sciences.

as their relata, and thus it reconciles the use of nomological statements in evolutionary theory with its populationist ontology.

In discussing the propensity interpretation, I have emphasized its role in induction. It is worthwhile here to similarly look at the inductive purport of the supervenience thesis: what kind of inductive reasoning does it support? The first thing to note is that these two theses are not rivals but rather play complementary roles, aiming to buttress the explanatory power of evolutionary models from different angles. The propensity interpretation takes care of the *diachronic* reasoning or prediction of evolutionary change: its goal is to show that an evolutionary model is "projectible" to the future and can be repeatedly applied for multiple generations to yield a significant evolutionary change. In this sense, it aims to ground an inference from now to then. The supervenience thesis, in contrast, focuses on the *synchronic* side of induction. It bestows selective explanations with generality to unify physically distinct or geographically distant populations under the same evolutionary model. This allows us to extrapolate the evolutionary dynamics observed in one population to another–to make an inference from here to there.

In the physical sciences, ampliative reasoning is guaranteed by the basic laws of a theory. The matter is more complicated in evolutionary biology, however, due first to the seemingly a priori nature of its fundamental principles and second to the lack of natural kinds. The propensity interpretation and the supervenience thesis were introduced by philosophers to address each of these problems, and to warrant diachronic induction and synchronic extrapolation of evolutionary models. They are, as it were, "the philosopher's stones" for turning the dry mathematical formulae into ampliative laws.

In fact, these magical stones were not mined in the land of the philosophy of biology, but rather were imported from the neighboring field of the philosophy of mind, where they had been used to make sense of mental states and justify psychological explanations. While propensity or dispositional properties were introduced to paraphrase mental states such as believing or knowing in behavioristic terms (Ryle, 1949), supervenience served as a weapon in the antireductionist arguments which refused to reduce the mental to the physical (Putnam, 1975; Fodor, 1974; Kim, 1992). Philosophers of biology have adopted these existing resources to solve their own problem. The patch was well accepted, with the two interpretations of fitness coming to define one of the "received views" in the philosophy of biology. Under this paradigm the case of the tautology problem initiated by Smart and other skeptics was eventually closed by the end of the 1980s, and the philosophy of biology came to flourish as a major subfield of philosophy of science toward the end of the century.

Happily ever after? Unfortunately not. Within a few decades, the case was reopened by a group of philosophers who questioned the received view and urged a reconsideration of the empirical and causal nature of evolutionary theory. The next section turns to this new skepticism.

3 The Statisticalist Controversy

As suggested in the end of the previous section, the philosophical analyses of fitness in the works of Beatty, Brandon, Mills, Sober, and to some extent Rosenberg in the 1980s not only offered a solution to a particular set of conceptual problems but also laid out the standard metascientific framework or "paradigm" under which evolutionary theory was characterized, construed, and analyzed in the subsequent philosophical literature. The crux of this received view consisted in the distinction between a theory and its application. Although mathematical models of evolution may be a priori, they receive their empirical contents when applied to actual evolutionary phenomena with the aid of source laws and proper interpretations of the concepts or quantities in the models. Philosophical interpretations, especially that of fitness, thus play an essential role in *making* evolutionary theory empirical.

From the beginning of this century, however, the received view came under vigorous attack by a group of philosophers called *statisticalists* (e.g., Matthen and Ariew, 2002; Walsh et al., 2002, 2017). The dispute primary concerns the nature of the "empirical contents" that evolutionary theory is supposed to deal with. Whereas the received view conceives of evolutionary theory as a study of the *causes* of evolutionary change, the critics retort that the theory, properly understood, concerns just statistical trends of population change with no regard to their causes – the empirical content of evolutionary theory is therefore of a purely statistical nature. According to this new view, population genetics is essentially an applied statistics and its model applies to actual populations through statistical estimation, whence the name *statisticalism*.

As an alternative interpretation of the role and nature of mathematical reasoning in evolutionary biology, the statisticalist contention hinges on the way formal models or principles of evolution are applied to actual phenomena, or in Sober's parlance, the nature of the source laws. The statisticalists' argument is twofold, consisting of a negative claim that design analysis, conceived by the received view as one of its source laws, falls short of fulfilling its job of estimating the parameters of evolutionary models, and a positive thesis that these parameters can be estimated only via statistical census. This section reviews these claims in turn, with a view to assessing their implications for the inductive nature of evolutionary theory.

3.1 The Statisticalist Criticism

In the received view, the source laws play a pivotal role in anchoring mathematical models of evolution to an empirical ground by providing numerical estimates of their parameters. We saw two kinds of such laws in the previous section, namely the statistical estimation from observed offspring counts and the design analysis of organismal morphology in given environmental condition. Put the statistical analysis aside for now and let us focus on the latter: what exactly is design analysis, and how does it serve as a source law?

The main task of design analysis is to identify and evaluate salient *causal factors* that contribute to fitness; that is, it aims to determine, among various characteristics possessed by target organisms, features that causally affect the chance of survival or the number of offspring. If you are a grazing animal living in open savannah and you can run fast, you will have a better chance of escaping from predators. Fleetness is thus a "good design" in this environment, a salient causal factor that positively contributes to the fitness of its bearers. So far, so good. But identifying positive causes of fitness is only the first step in the fitness estimation: one also has to numerically assess *the extent to which* such causes affect fitness. How do differences in speed relate to differences in fitness? Suppose the typical predators in the savannah can run up to 60 km/h. Then one might well infer that a particular design of legs that allows bearers to run at 70 km/h is more advantageous than one that permits only 50 km/h, but not so much more than the one that permits 65 km/h, and so on.

Can we continue on like this to obtain the desired outcome, that is, the abstract measure of fitness required by population genetics models to predict evolutionary change? Statisticalists think not (Matthen and Ariew 2002; Pigliucci and Kaplan 2006; see also Rosenberg 1982 for a similar remark). What can be obtained from such procedures is a bunch of pairwise comparisons, which might tell us which organism is more advantaged than another, but never to what extent.[8] Moreover, because each causal factor is qualitatively different, results of design analyses do not add up. Even if we determine fleetness and immunity as positive causes of fitness, we have no idea about how these two factor combine to yield, say, the fitness of a fast but sickly individual. But that is what is required by formal models of population genetics: "the *expected* rate of increase ... of a gene, a trait, or an organism's representation in future generations" (Matthen and Ariew, 2002, p. 56, their emphasis). This rate expresses not just which individual (or gene or

---

[8] In fact this assumption can be questioned (Otsuka, 2016a), but I will ignore this point here for the sake of argument.

trait) fares better than another, but also the extent to which they will increase or decrease in the next generation, with specific numerical values.

What statisticalists point to here is a difference in *scale*. Design analysis, according to statisticalists, is like grading an essay and gives only an *ordinal scale* or rank ordered according to a certain criterion (A, B$^+$, B, etc.). From such a grading system you can see for any given student whether she or he did better or worse than any other student, but it does not tell you whether getting two Bs is equivalent to one A and one C, or whether B$^+$ is as far away from B as B$^-$ is from C$^+$. In contrast, the fitness value required by mathematical formulae must, like height or weight, have a *ratio scale* in which each interval is meaningful and has a definite value. The difference between a height of 170 cm and one of 160 cm is not just 10 cm but also twice the difference between 155 cm and 150 cm. Similar quantitative comparisons must be meaningful with respect to fitness in order to quantitatively calculate the extent to which one type will increase or decrease its frequency against the others. The claim of Matthen and Ariew (2002) can therefore be understood as stating that the outcomes of design analysis and the fitness parameter are on different scales, and that this fundamental gap disqualifies design analysis as a source law.

If design analysis falls short of its goal, the only option left to estimate fitness is by statistical estimation. That's just the way it is, say statisticalists. "The overall fitness values demanded by consequence laws must be estimated statistically, that is, by looking at actual values for number of offspring, and using these actual values to estimate expected values and other statistical quantities" (Matthen and Ariew 2002, p. 67; see also Rosenberg 1982). As I understand it, what they suggest in this passage is the statistical method of *regression*: one measures actual offspring counts W and target trait Z, and by regressing the former on the latter one obtains a quantitative relationship between them, and in particular, an estimate of the expected fitness conditional of the given trait, $E(W|Z)$. This is, as we have seen in the previous section, exactly what the propensity interpretation of fitness is. And as I suggested there, such a statistical estimation requires no knowledge about causality. All you need are data, that is, measurements of fitness and the trait under study, and then relevant statistical techniques take care of converting sample statistics into regression estimates and evaluating their reliability. Knowledge about design or causes, that is, whether or how the trait affects fitness, plays no role here. On this ground, statisticalists conclude that the proper and only possible form of source law is *statistical estimation of population parameters from finite samples*, and nothing else.

3.2 Statisticalist Interpretation of Evolutionary Theory

Source law, correctly construed, deals only with statistical, rather than causal, relationships. From this seemingly minor modification to the received view, Matthen and Ariew derive the sweeping conclusion that evolutionary theory is by nature a statistical, not a causal, theory. In union with the related works published around the same period (e.g., Walsh et al., 2002; Matthen and Ariew, 2005) this view came to define an alternative to the received view, often called the *statisticalist* interpretation of evolutionary theory. As the name suggests, it portrays evolutionary theory as a purely statistical theory, as a study not of causes but just of statistical trends of population change. But why does such a conclusion follow? To see the statisticalists' logic, recall that in the received view, the source laws along with the proper interpretations of fitness take the whole responsibility of giving empirical contents to mathematical models of evolution (Fig. 2.1) – they are what *make* evolutionary formulae empirical. Hence if these formal models or conse-quence laws of evolution have something to say about causes of evolution, there must be a corresponding source law that takes up the relevant causal facts about an evolving population and relates them to the consequence law. But closer inspection of design analysis revealed that causality has no bearing on the numerical estima-tion of fitness. If no causal assumption goes in, no causal explanation comes out. Evolutionary theory therefore is concerned only with statistical trends and "explains the changes in the statistical structure of a population by appeal to statistical phenomena," or so statisticalists argue (Walsh et al., 2002, p. 471).

The statisticalist thesis has invoked strong opposition from those who hold that key concepts in evolutionary theory such as selection and drift do represent causal processes and provide causal explanations rather than just statistical

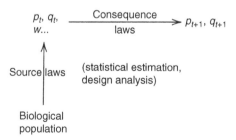

Figure 2.1 Sober's picture of evolutionary theory. The source laws, through statistical estimation and design analysis (discussed below), "encode" a biological population into quantitative conditions and parameters such as gene frequencies p, q or fitness W, from which consequence laws calculate evolutionary change from generation t to $t + 1$.

summaries (e.g., Reisman and Forber, 2004; Bouchard and Rosenberg, 2005; Millstein, 2006; Millstein et al., 2009; Ramsey and Brandon, 2007; Ramsey, 2013, 2016). Here in this Element, however, I limit myself to the examination of the statisticalists' claim itself without going into the debate,[9] because their view on the explanatory structure of evolutionary theory presents an alternative approach to our main question on the applicability of mathematical formulae to empirical evolution.

Let us start by unpacking the claim that evolutionary models explain "statistical trends" of populations: how do such explanations take place? Like the received view, statisticalism holds that the general laws of evolution are a priori mathematical theorems, such as the Price equation or Li's theorem,[10] which by themselves make no empirical assumptions about phenotype, fitness, inheritance, population structure, and so on. When they are applied to actual evolutionary phenomena, however, these biological details must be spelled out, and this is how evolutionary explanations take place, according to statisticalists; that is, a particular episode of evolutionary change is explained when it is subsumed under an abstract formula through the specification of its theoretical conditions with concrete information of the population and environment under study. The overall idea is thus quite similar to the received view's two-part framework of mathematical deduction (consequence laws) and empirical application (source laws), but there are two important differences. First, and this is what we have just seen, the specification of theoretical premises requires no information about causes, but only statistical data. In the case of the Price equation $\Delta \overline{Z} = Cov(W,Z)/\overline{W}$, to be specified are the trait Z and its covariance with fitness as well as the mean fitness value, but not the causal relationship between them. Second, there are different levels of specification, giving rise to models of different generalities. If a model covers every concrete detail of a population, it will provide the most specific and detailed explanation of that population. But if one chooses to specify just a system of inheritance and leave out other factors, the resulting "semi-abstract" model would apply to all populations with that particular inheritance system. In this way, one can obtain various models with different levels of abstraction and scopes, and for this reason Matthen and Ariew call their view the *hierarchical realization scheme* (Matthen and Ariew, 2002, 2005).

According to this hierarchical scheme, evolutionary explanations take place by subsuming concrete populations under general patterns, which are further

---

[9] My review of the debate can be found in Otsuka (2016a).

[10] Li's theorem (Li, 1955) is a special case of the Price equation where phenotype Z is fitness W itself, and it states that the change in fitness is proportional to its variance such that $\Delta \overline{W} = \mathrm{Var}(W)/\overline{W}$.

subsumed by more general patterns, continuing up to the most abstract formulae of evolution expressed in pure mathematical forms. The explanatory strategy that proceeds through such progressive subsumptions of particular events under more general patterns exemplifies the *unificationist view* of explanation (Friedman, 1974; Kitcher, 1981). On this view, the explanatory power of a scientific theory consists in its ability to unify different phenomena under a few key principles. The Galilean physics, for example, improved upon the preceding Aristotelian theory in that it provides the unified treatment of both celestial and terrestrial bodily movements. Likewise, one may think the significance of Darwin's theory resides in its ability to deduce "endless forms most beautiful and most wonderful" with the single principle of descent with modification. The mathematical formulae of evolution extract the essence of this deductive process in the most abstract form (Morrison, 2000; Luque, 2017) to cover even the "evolution" of nonbiological entities. Population genetics explains frequency changes of particular populations by subsuming them under such a unified scheme, just as Darwin explained various forms of life as instances of his principle of adaptive evolution, albeit in a more precise and systematic fashion.

The unificationist approach also illustrates how an evolutionary model generalizes, for generality is just another name for unification. One model is generalized to multiple populations insofar as it subsumes and unifies evolutionary dynamics of these populations; and this is when, according to statisticalists, the model specifies the least empirical conditions common to all the target phenomena. Hence by starting with the most specific model that applies to just one population, say a population of Galapagos finches on one island in a particular year, one can generalize this model by "abstracting away" some of its conditions. One may, for instance, drop the environmental condition specific to that particular year to obtain a more robust model of Galapagos finches; or remove physiological specifications altogether to get a general model of adaptive evolution under changing environment that applies to any species. Through such abstractions models come to "capture evolutionary phenomena in their full generality, as they apply equally to populations of *Latimeria* and *Laminaria*, bacteria colonies and Bactrian camels, and much more besides" (Walsh et al., 2017, p. 12). Showing such general and robust patterns or regularities that are insensitive to specific biological details is, according to statisticalists, the prime goal of evolutionary theory, and their hierarchical realization scheme captures this explanatory strategy.

3.3 Does Statisticalism Solve the Problem of Induction?

Now that we have seen the statisticalists' take on evolutionary theory, let us examine closely its implications for our question as to how mathematical models support inductive reasoning about evolutionary change. Can statistical-ists make sense of the inductive use of mathematical models of evolution, like the use of the population genetics formula (Eqn. 1.1) for predictions of future evolutionary trajectories that we saw in Section 1? Given that statisticalism, like the received view, holds the fundamental laws of evolution to be mathematical theorems (Matthen, 2009; Matthen and Ariew, 2009), it is pertinent to ask whether and how such a priori principles help the inductions and predictions made in evolutionary biology.

In the received view, a mathematical model of evolution is applied to a target population via the estimation of its parameters by design analysis or statistical estimation. But since statisticalists deny design analysis as a proper source law, the application now hinges just on statistics: that is, an evolutionary phenom-enon is subsumed and explained by some mathematical formula when its parameters are statistically estimated from data taken from the target popula-tion. In this sense, evolutionary explanations are just routine practices of applied statistics, that is, estimations of statistical parameters. But are they really so? Statistics is all about estimating a population (in the statistical sense) from samples. Suppose you want to buy a used car and decide to do some market research. You might, for example, want to know the average price of used cars as well as whether and how the price depends on the manufacturing year. For this purpose you collect some samples from the market, record the price W and the model year Z of each, and use these data to calculate the average price $\overline{W}$ and covariance $\mathrm{Cov}(W,Z)$. Statistics then helps you test whether your estimates reflect the real pattern in the market or just a peculiarity in your sample. On the face of it, estimating the parameters of evolutionary equations such as the Price equation looks no different from this, though in this case we are interested not in the price (no pun intended!) and year of cars but fitness W and average offspring phenotype Z' of organisms; that is, we record these values from individuals sampled from a population, and from this sample estimate the population covariance $\mathrm{Cov}(W,Z')$ and mean fitness $\overline{W}$. If the estimated covariance sig-nificantly deviates from zero, we can conclude that the population is indeed evolving. This seems to be a straightforward, routine statistical inference.

This, however, is not the end of the story. What we have just concluded is whether *this particular* population is evolving or not *now*, that is, between two observed generations. But that is not what interests us in most evolutionary studies: we rather want to *predict* the future evolutionary change or *extrapolate* the evolutionary dynamics of one population to another, whereas nothing in the above statistical estimation allows for such ampliative reasoning. Returning to the car example, your statistical estimate of the relationship between price and year is valid just for the population you sampled from. If you move to another state, you can no longer rely on your estimate, unless there is a strong reason that the used car markets in your new and old home states are mostly the same.

In evolution, however, there are strong reasons that two populations, or the same population at different generations, are *not* the same. For one thing, evolution changes phenotypic or genetic frequencies, and thus the statistical structure of the population itself. If a population evolves from generation t_1 to t_2, then by definition the generations are not the same statistical populations, and thus there is no reason, at least within statistics, for them to have the same statistical parameters. Hence if we want to "apply" the Price equation to this new generation, we need to reiterate the whole estimation procedure and sample anew, and likewise for any subsequent generation. But that means "evolutionary explanations" as conceived by statisticalists have no predictive ability – what they can do is merely describe evolutionary response at each generation *post factum* (van Veelen, 2005; van Veelen et al., 2012). True, inferential statistics involves predictions of unobserved samples from observed ones, but such inferences must be made within a fixed population and should be distinguished from predictions of evolutionary responses that necessarily go beyond a particular population here and now. Statistics is an art of inferring the whole from its parts, but it does not tell you anything beyond a particular population or how that specific population changes. Theoretical justifications for such extra-population reasoning, therefore, must come from somewhere else than statistics or the statistical structure of the target population.[11]

In the previous section I noted that the received view's propensity interpretation and source laws served as the uniformity of nature needed to ground evolutionary induction. A successful induction from one population to another presupposes the existence of a certain uniformity that is shared by both populations and connects an observation of the former to a prediction of the latter. The propensity fitness that supervenes on organismal designs was expected to play such an anchoring role, authorizing an extrapolation of the same model among populations or species that exhibit the same or functionally equivalent design.

---

[11] Note that the contention here is *not* about the so-called dynamic (in)sufficiency of an evolutionary model, which concerns a model's recursive ability (Lewontin, 1974). If a transition function can be repeatedly applied to its outcome to calculate an evolutionary consequence of arbitrary future, such that $X_n = \underbrace{f \circ \ldots \circ f}_{n} (X_1)$ for any generation $n \geq 1$, a model is said to be dynamically sufficient; while dynamically insufficient models can be applied only one or limited times. In our examples, the one-locus population genetics model (Eqn. 1.1) is dynamically sufficient. Here the claim is not that the Price equation is dynamically insufficient, but rather that it does not make *any* prediction, even that of one generation after. Dynamical (in)sufficiency is a property of predictive models, and thus attributing it to the Price equation is a category mistake (Frank, 1995; van Veelen, 2005; Luque, 2017).

Statisticalists, however, denied the role of design analysis and replaced it with statistical estimation, which means the required uniformity of nature must be sought in a statistical population and its parameters. At first sight this looks just like how standard inferential statistics works, where the assumption of a statistical population serves as the uniform nature that warrants inductive inferences from observed to unobserved samples. But in contrast to statistical inferences that assume a fixed population and are made within that population, evolutionary studies are concerned with changes of the very population or inferences between populations, and on such population dynamics or extrapolations the conventional statistical methods have nothing to say. Hence, if statistical properties of a population are all there is to know, inductive reasoning about evolution becomes a metaphysical impossibility.

Some statisticalists seem to be aware of this difficulty and for this reason propose to drop prediction altogether from the tasks of evolutionary biology (Pigliucci and Kaplan, 2006, p. 61). Pigliucci and Kaplan dismiss the insistence on quantitative predictions of evolutionary trajectories as an off-the-shelf ideal uncritically borrowed from the standards of the physical sciences, and claim that evolutionary biologists should focus more on the elucidation of a complex causal nexus among genetic, phenetic, and environmental factors operative in each historical case. Although it is certainly true that evolutionary biology comprises much more than just predicting population changes, giving it up altogether as "physics envy" is too quick a move, for it would put into question not just the whole of population genetics but also the theoretical status of evolutionary theory as an inductive science. As we saw in Section 1, the ability to make predictions and extrapolations marks the essential feature of the modern evolutionary theory that saved Darwin's theory of natural selection and distinguished it from the descriptive natural history of the eighteenth century, and population genetics was expected to provide the logical basis or law for such inductive reasoning. If this attempt turns out to be impossible or infeasible as Pigliucci and Kaplan suggest, the whole of evolutionary theory would be reduced to a heap of empirical observations lacking its own unity or principles, just as Smart (1959) argued.

Our quest for the ampliative basis of evolutionary theory, therefore, is brought back to square one. We started with the question of why Darwin's principle of natural selection, expressed as mathematical formulae, can ever predict and explain evolutionary phenomena. The received view sought to answer this question with the interpretation and application of these formulae and proposed two kinds of source laws, design analysis and statistical estimation. The statisticalists then dismissed the former and kept the latter as the only possible way to apply evolutionary models. The above discussion, however, revealed that what such an "application" achieves is at most a description of the present evolutionary change, which falls far short of a prediction or explanation of future or unobserved trajectories. After this philosophical tour, we still don't understand why mathematics can say anything new about evolution.

So we have to begin anew. But from where? As suggested in the beginning of the previous section, the past philosophical investigations have taken the whole issue as a semantic problem and sought a correct interpretation of mathematical formulae that would make them empirical. But is it the only, or even an effective, approach to understand the role of mathematics in evolutionary theory? Why, in the first place, does giving a meaning to an abstract formula make it empirical? The next section steps back a little and examines the methodological suppositions shared by both the received view and statisticalism, with a view toward hinting at an alternative approach to tackle the problem.

4 Beyond Dualism

The previous sections have reviewed two major philosophical interpretations of evolutionary theory. The two camps – the received view and statisticalism – hold contradicting views on how abstract mathematical models relate to actual evolutionary phenomena, and the disagreement has provoked a significant debate over the past few decades that still continues today (e.g., Otsuka, 2016a; Walsh et al., 2017). But despite the apparent discrepancy, from a deeper, metaphilosophical perspective they share the same conceptual as well as methodological frameworks. What is common in these views is a kind of dualism, which portrays evolutionary theory as consisting of two parts, namely (i) the formulation of mathematical models or laws of evolutionary dynamics through a priori deductive reasoning, and (ii) the application of these formal models to concrete populations through a posteriori estimation of relevant parameters. In the received view, these two parts are represented by consequence laws and source laws, respectively, while in statisticalism they correspond to statistical models and their estimation through census. In this dualistic landscape, the problem of understanding the role of mathematics in evolutionary explanations boils down to the question of whether and how the two separated realms can be reconciled. The connection is established, it was hoped, by a proper interpretation that maps mathematical concepts or parameters to concrete, empirical phenomena. The dualistic framework has thus put interpretation as the primary business of philosophers. It is in this methodological framework that the received view and statisticalism have tried to ensure the empirical relevance and inductive power of mathematical models, by interpreting fitness as a supervenient propensity or a statistical parameter, respectively.

From a broader perspective, this theory/application dualism, along with the interpretative methodology, can be tracked down to the metascientific view of the logical empiricists. The central tenet of logical empiricism is the analytic–synthetic distinction, namely the distinction between statements that are true or false just by their logical form or the meaning of the terms they comprise, and those whose truth value depends on an empirical state of affairs. For the empiricists, the whole of mathematics belongs to the realm of the analytic a priori, a kind of logical tautology. This immediately raises a conundrum as to how mathematical scientific theories like the theory of relativity can say anything about the actual course of nature. This problem is solved, it was believed, by relating the formal and deductive components of a theory to the empirical ground via *correspondence rules*.

> [A] deductive theory can be built up by combining [theoretical] postulates with logic and mathematics, but the result is an abstract deductive system in which the theoretical terms have not even a partial interpretation ... To become a scientific theory, its descriptive terms must be interpreted, at least partially. This means that its terms must be given empirical meanings, which is done, of course, by correspondence rules that connect its primitive terms with aspects of the physical world. (Carnap, 1966, p. 267)

For instance, Euclid's proof that the three interior angles of a triangle always add to 180° doesn't say anything about an actual figure on a piece of paper unless the term "triangle" is interpreted to denote that particular figure. Likewise, Einstein's relativity theory, construed as a formal study on Lorentzian manifolds, becomes an empirical hypothesis with testable predictions only if those mathematical objects are interpreted as actual space-time. In both cases the correspondence rules function as a bridge between abstract mathematics and empirical observations, which provides analytic statements with empirical contents and allows them to make synthetic predictions. In other words, it is these interpretive rules that *make* a deductive system into an empirical scientific theory.

This is exactly the strategy taken by philosophers of biology to make sense of the applicability of mathematical models to evolution. They construed population genetics (consequence laws) as an "abstract deductive system" and sought to anchor it to an empirical ground via "partial interpretations" of theoretical terms, most notably fitness.[12] The propensity interpretation and

---

[12] In logical empiricism, interpretations serve to translate sentences containing theoretical terms (which, like "electron" or "temperature", lack direct observational contents) to those involving only observational terms. Since the meanings of theoretical terms are not exhausted by finite observations (the term "electron" refers to not just past and present experimental observations but also all future ones not yet observed), such an interpretation is always *partial*. The same holds true with the propensity fitness, which cannot be totally determined by actual offspring counts, but only partially inferred via statistical estimation.

supervenience thesis are nothing but such correspondence rules that connect fitness with relevant biological features and furnish evolutionary theory with due empirical purports. It is also no wonder that interpretation takes on a primary significance for the metascientific investigation under this framework. The formal theory of evolution, construed as an applied mathematics, is empirically transparent as it were, and its nature as a scientific theory is determined entirely by the nature of its correspondence rules. It is this (mostly implicit) assumption that the statisticalist controversy has hinged upon for the correct interpretation and estimation procedure of fitness. Estimation is what relates a theoretical quantity to actual observation – that is, a correspondence rule. Hence in order for evolutionary theory to say anything about the causes of evolution, its estimation process must reflect some causal features of organisms. If, on the other hand, it concerns only statistical properties of a population, then the theory would be all about "statistical phenomena," as statisticalists argue (Walsh et al., 2002).

One may thus view the quarter-century-long struggle in the philosophy of biology to understand the mathematical nature of evolutionary theory as a deployment of the Carnapian program. Some may find this incongruous, given that many philosophers of biology were outright dissidents of the logical empiricist tradition (e.g., Brandon, 1981; Beatty, 1981; Rosenberg, 1985). Most of their complaints, however, are targeted at the reductionist slant of logical empiricism and its alleged emphasis on the essential role of universal laws in scientific theorizing, as typified by Smart's criticism we saw in Section 2. In contrast, the analytic–synthetic distinction and the notion of correspondence rules for bridging this gap were mostly untouched, or even implicitly accommodated by philosophers of biology for the service of their own task of analyzing the nature of evolutionary theory.

As is well known, however, this distinction between analytic/mathematical statements and synthetic/empirical ones, along with the empiricist view of scientific theories, came to be questioned in the second half of the twentieth century, mostly due to Quine's "Two Dogmas of Empiricism" (Quine, 1951). In this seminal paper Quine attacked two hallmarks of analyticity and claimed that both fail to distinguish analytic from synthetic statements. First, analytic statements are often defined to be "true by their meaning" – "all bachelors are unmarried" is said to be analytic because it is true by the meaning of the word "bachelors." Such judgments are based on the notion of synonymy, but Quine argued that one cannot establish the sameness of meaning without presupposing the very notion of analyticity – hence saying that some statement is analytic because it is true by its meaning is just circular and tells nothing about the nature or criterion of analyticity. The second hallmark of analytic statements is their immunity to empirical refutation, that is, they "stand come what may." Quine criticizes this idea for assuming too simplistic a picture of verification where each proposition is confirmed or disconfirmed by experience independently from the rest of beliefs. Against this reductionist conception of verification he offered a holistic picture where our beliefs are so tightly intertwined with each other that they "face the tribunal of sense experience not individually but only as a corporate body," with there being no unique rule that dictates which part of the whole web of beliefs should be conserved or revised when a discrepancy arises. So even propositions like "the earth is flat" can be made immune to falsification by introducing auxiliary hypotheses or resorting to a hallucination, while logical laws such as "it is not the case that P and not P" could be discarded if necessary (as dialetheists would argue). Although it is true that in most cases we tend to be more reluctant to give up logical or mathematical truths than other beliefs, it is simply because they are situated at the core of our knowledge so that their revision leads to a large-scale rearrangement of our entire belief system. Hence all sentences are on a continuous spectrum that scales revisibility against adverse evidence, with no objective boundary that distinguishes between the revisable and the unrevisable, or between synthetic and analytic propositions.

If the synthetic/analytic distinction is untenable, or at least not as unproblematic as Carnap had believed, the whole premise of the dualistic approach that a scientific theory can be analyzed into mathematical and empirical parts loses its justification. Science does depend on deduction and observation, "but this duality is not significantly traceable into the statements of science taken one by one" (Quine, 1951, p. 39); that is, a scientific theory cannot be partitioned into analytic components and synthetic ones. It rather forms a contiguous network, its theoretical assumptions around the center and its relation to experience along the edges. If that is the case, the task of a philosophical analysis is to elucidate, without drawing an arbitrary line between a priori and a posteriori, the structure and organization of such a system of beliefs, and its functioning in predicting and explaining of our experience. But how? Unfortunately Quine himself did not engage in a detailed analysis of any particular scientific theory, nor did he say much about how such a project should be carried out, but we can take a cue from another ardent critic of the logical empiricist tradition, Patrick Suppes.

Like Quine, Suppes (1967, 2002) criticizes the dualistic approach as "far too simple" to capture the complexity and sophistication of scientific theorizing and practice. In particular, he complains that identifying all the theoretical bodies of science with simple logical calculus overlooks the rich and diverse nature of formal theories and leads to a false dichotomy of mathematical versus empirical sciences. Given that most scientific theories involve advanced mathematical operations that go far beyond elementary logical calculus, and that the requisite mathematical operations differ from one theory to another, an analysis of a scientific theory must begin with identifying the formal and mathematical background upon which its theoretical body is constructed;[13] that is, rather than lumping together all theoretical components as "mathematics," we should carefully dissect the assumptions of a scientific theory and determine what kind of mathematical concepts are used to express its hypotheses, predictions, and conclusions. Indeed, the rich diversity of today's mathematical theories makes it hardly informative to call some statement a "mathematical truth." Is it a theorem of group theory, differential geometry, or something else? In the standard mathematical presentation, each of these mathematical theories is defined by a set of axioms, which determines the nature and content of the theory. At the same time these theories are not unrelated to each other but often show logical dependencies: probability theory, for instance, is built upon the basis of measure theory, which in turn assumes the theories of reals, topological spaces, sets, and so on. Axiomatization, or what Suppes calls the definition of a *set-theoretic predicate*, serves to clarify the nature of mathematical theories and the logical relationships among them. Suppes argues that the same holds true of scientific theories: just like mathematical theories, a scientific theory is built upon a certain mathematical background and its own set of premises, so making these explicit in an axiomatic form would clarify the formal nature of the theory. Newtonian mechanics, for instance, stands upon its celebrated laws of motion as well as mathematical tools like differential calculus. An axiomatization makes an "inventory list" of these assumptions and helps us sort out the proper and specific assumptions of the theory from those "borrowed" from other mathematical theories. With this prospect Suppes tried to axiomatize Newtonian mechanics and formal learning theory and construct them on top of the requisite mathematical theories, just as complex mathematical theories are built upon more basic ones (Mckinsey et al., 1953; Suppes, 2002).

The Suppesian approach emphasizes the continuity between mathematical and scientific theories: "there is no theoretical way of drawing a sharp distinction between a piece of pure mathematics and a piece of theoretical science" (Suppes, 2002, p. 33). With regard to our inquiry, this continuity implies that mathematical and scientific theories are to be analyzed in the same way, through

---

[13] Another, no less important part of Suppes's claim is his view on models of experiments, data, and measurement, which addresses the question as to how a formally characterized scientific theory relates to empirical data (Suppes 1962, 2002). Though important, this topic cannot be covered in this Element for want of space.

an exposition and elucidation of a set of premises or axioms that define the theory in question. In mathematics, axioms serve as both building blocks and a hallmark that distinguishes one theory from others: Kolmogorov's axioms, for instance, underlie the whole contents of probability theory and at the same time distinguish it from other variants of measure theory. Now, if a scientific theory is constructed upon and in continuity with background mathematical theories, as Suppes claims, it can also be analyzed in the same fashion, namely by identifying its proper premises and tracking how these premises lead to its formulae, concepts, and theorems. In his axiomatic reconstruction of classical particle mechanics, such basic axioms include Newton's second law of motion ($F = ma$) and specifications of the nature of space, time, and mass, which together allow for the derivation of the body of Newtonian physics including the principle of inertia, the composition of forces, the conservation of energy, and so on. It is these axioms or premises that define Newton's theory and distinguish it from other mathematical or scientific theories. The Suppesian approach thus focuses on the internal logical structure of a scientific theory and in this respect is in contrast to the dualistic approach that looks to extratheoretic relationships such as interpretations or applications.

Can we apply the Suppesian strategy to our investigation of the theoretical and mathematical nature of evolutionary theory? To axiomatize evolutionary theory or even a part of it appears to be a daunting task.[14] But axiomatization, ideal though it may be in terms of rigorousness and thoroughness, is not the only way to study the formal structure of a scientific theory. If one's interest is in specific aspects of a theory, it makes perfect sense to focus on the relevant formulae and examine the assumptions and mathematical background requisite to derive these results. In the context of our investigation, the question in focus is the mathematical basis of population genetics models that allow for inductive reasoning of evolutionary change. Many of the philosophers of biology discussed so far did not doubt that these models are built within probability theory and on this basis concluded that the formal theory of evolution is just applied mathematics (Brandon and Beatty, 1984; Rosenberg, 1985; Matthen and Ariew, 2009). On the face of it, such a presupposition seems to be justified by the oft-made claim that various evolutionary models are "derivable" from the Price equation (e.g., Frank, 2012; Queller, 2017). The fact, however, that a formula is derived from some probabilistic formula does not entail that it is derived *within* probability theory, for the derivation may make use of other assumptions that are simply not in probability theory. The Suppesian approach to evolutionary

---

[14] But see Williams (1970), Magalhães and Krause (2001), and Grafen (2014) for some attempts to axiomatize evolutionary theory. In particular Magalhães and Krause explicitly adopt the Suppesian approach to define a set-theoretic predicate of population genetics theory.

theory must identify the nature of these additional assumptions, if any, and a formal framework that is powerful enough to accommodate the whole derivation process. That may fall short of a full-fledged axiomatization, but it will nevertheless shed some light on the formal nature of evolutionary theory.

To recap, this section contrasted the traditional, dualistic approach with the Quine-Suppes approach to understanding the mathematical nature of evolutionary theory. Despite their apparent antagonism, the existing philosophical approaches to the role of mathematics in evolutionary theory we have considered, the received view and statisticalism alike, tacitly adopted the positivist program and tried to anchor a priori mathematical formulae on an empirical ground by providing proper interpretations or correspondence rules. Such a sharp distinction between analytic and synthetic components came to be questioned by Quine, who treated a scientific theory as a holistic network over various kinds of beliefs, mathematical and nonmathematical, that function together to achieve given epistemic goals. The primary task of a metascientific analysis in this view is then to delineate the relevant assumptions and the logical links between them. A rigorous axiomatization or definition of Suppesian set-theoretic predicates would be one way to achieve this task; but even without going that far, elucidating the formal assumptions and the process of derivation should shed no less light on how the theory and practice of evolutionary biology depend upon its mathematical principles. Indeed, so far we haven't said much about what these "evolutionary principles" are exactly, let alone where they come from. The next section takes on these questions, focusing more on the internal logical structure rather than external interpretations of population genetics models.

5 Causal Foundations of Evolutionary Theory

According to Suppes, a formal theory, scientific or mathematical, is best characterized by its axioms. They serve as building blocks of the theory and determine what it can do or prove. In this section we explore the path suggested by Suppes and try to identify the theoretical premises of the mathematical formulae that underlie the practice of evolutionary inferences. Of particular interest are predictions of adaptive evolution, such as a radiation of particular genes or phenotypic change in response to a given selective pressure (Section 1). In order to make sense of such inductive reasoning, past philosophical investigations started with a priori formulae and tried to *make* them ampliative by way of interpretations or applications. But as we saw in Section 3 this approach led us to an impasse: mathematical identities such as the Price equation, however interpreted or estimated, never give information not contained in data, and for this reason they are seldom if ever used for actual scientific or pragmatic purposes like predicting the evolution of antibiotic-resistant bacteria or the response of a wild population to changes in climate. We now take the opposite approach: start with formulae that are actually used by researchers for predicting evolutionary response and examine the theoretical assumptions of these models. This will clarify the "inductive basis" of evolutionary reasoning and the conditions under which these mathematical formulae can make correct predictions about actual evolutionary change.

5.1 Causal Foundations of Evolutionary Theory

In Section 1 we saw Lewontin's schematic representation of mathematical genetics as a study of a population trajectory in the genotypic and phenotypic spaces (Fig. 1.2). As noted there, most population genetics models are built as transition functions that track evolutionary changes in either of these two abstract spaces. The first step to examine the theoretical assumptions of evolutionary models, then, is to take some such transition functions and ask how they are constructed. We have already seen the simple one-locus population genetics model:

$$\Delta p = \frac{pq[p(w_{AA} - w_{Aa}) + q(w_{Aa} - w_{aa})]}{p^2 w_{AA} + 2pq w_{Aa} + q^2 w_{aa}}. \tag{5.1}$$

As we saw in Section 1, this equation enables us to calculate a change in the gene frequency, Δp, from the current genetic frequencies p, q, and the genotype fitness w, and through recursive applications to predict the genetic distribution of a population at any future generation (Fig. 1.3).

There are also models for phenotypic evolution. A typical example is the *breeder's equation*, which gives the change in the mean phenotype $\Delta \bar{Z}$ in response to selection:

$$\Delta \bar{Z} = Sh^2 \tag{5.2}$$

where S is called the *selection differential* and h^2 the *heritability*. The selection differential S represents the strength of selection and is defined as the change in the mean phenotypic value of the parental population before and after selection (but before reproduction). Imagine a local population of house sparrows with an average weight of 20g. The population was struck by a severe snowstorm on a winter day that killed smaller individuals, with the average weight of the survivors being 23g. Then the selection differential for this selective episode is $S = 23 - 20 = 3$g. The heritability h^2, on the other hand, represents how much of this difference caused by selection is passed down to the offspring generation.

Sparrows big enough to survive the storm are expected to beget bigger-than-average offspring, but to what extent? According to the theory, this depends on the extent to which the variation of weight is determined from the variation of underlying genes, that is, the proportion of genetic variance within the total phenotypic variance.[15] The heritability h^2 expresses this degree of "genetic determination" of a phenotypic variance in the range between zero and unity. Hence if the heritability of body weight of the above sparrows is 0.2, the breeder's equation predicts the mean body weight will increase by $0.2 \times (23 - 20) = 0.6$g between the two generations.

Given that the one-locus population genetics model and the breeder's equation respectively represent transition functions in the genotypic and phenotypic spaces, our next task is to identify the theoretical premises on which these formulae stand. As Léwontin (1974) observed, successful models must adequately capture the four transition steps $T_1 \sim T_4$ that together constitute evolutionary changes in his scheme (Fig. 1.2). Just to recall, these steps were: (T_1) developmental mapping from genotype into adult phenotype; (T_2) phenotypic transitions by selection and other evolutionary factors; (T_3) gamete production by surviving individuals; and (T_4) fertilization and formation of new genotypes. Even a model that appears to be concerned with evolutionary changes in just one dimension, either genotypic or phenotypic, tacitly encodes all these steps in terms of its functional form and parameters. In the case of the breeder's equation, selection differential S parameterizes the transition T_2 in the phenotypic space, whereas heritability h^2 takes care of all the remaining steps T_3, T_4, and T_1. This suggests that evolutionary transition functions can be obtained by specifying these transition steps.

But how? The transitions $T_1 \sim T_4$ summarize causal processes of given life stages: development, selection, mating, and reproduction each mark particular sequences of causal processes that together constitute an organism's life cycle. Development, for instance, is nothing but the causal relationship from genes to phenotype, whereas phenotypic transitions by selection result from a phenotype's causal contribution to survival and reproduction, and likewise for other steps. We can thus expect that these causal relationships underlie the evolutionary transition functions and serve as the basis for their theoretical construction.

This expectation is borne out by *causal graph theory*, a formal framework to deal with the systematic relationships between causality and probability

---

[15] More precisely, the (narrow sense) heritability is the proportion of a phenotypic variation that is accounted for by the additive effect of genes. For an excellent review of the concept and its complicated history, see Tabery (2014).

(Wright, 1934; Spirtes et al., 1993; Pearl, 2000). As the name suggests, the theory expresses causal structures with directed graphs – sets of variables connected by arrows representing causal influence – and studies how they generate and constrain probability distributions over the variables. Using this theoretical apparatus, one can derive an evolutionary transition function from causal assumptions in three steps (Frank, 1997; Rice, 2004; Otsuka, 2016b). The first step is to identify the causal structure underlying the transition steps $T_1 \sim T_4$ and represent it in the form of a *causal graph* over relevant variables including fitness, phenotype, and genes of parents and offspring. Next, identified causal relationships are parameterized by *structural equations* that specify the quantitative relationships between the causes and the effects. On this causal model, the axioms and rules of causal graph theory generate a probability distribution from which an evolutionary transition function is calculated. These derivation steps are illustrated in Box 1 for the case of the breeder's equation.

A causal model like Fig. 5.1 encodes a set of causal assumptions in the formal parlance of causal graph theory. It asserts, for example, that the phenotype under study affects fitness in the way specified by the graph and the structural equations. The sketch of derivation demonstrates that evolutionary transition functions are based on such causal assumptions. It also implies that the correctness of evolutionary models crucially depends upon the veracity of these assumptions, so that the breeder's equation fails to obtain if the causal structure of a population under study deviates from Fig. 5.1. To see this, suppose that instead of phenotype Z causing fitness W as in the graph, they are *confounded* by some unknown factor U so that $W \leftarrow U \rightarrow Z$, where U may be some environmental factor or another trait. In this case no evolutionary response follows (i.e., $\Delta \overline{Z} = 0$) even though one would still observe nonzero heritability and selection differential due to a spurious correlation between Z and W. That is, the breeder's equation, or any evolutionary model for that matter, does not hold if its causal assumptions are violated, which attests the importance of checking causal assumptions in the application of evolutionary models (e.g., Rausher, 1992; Morrissey et al., 2010).

In addition to this practical moral, confounding and the resulting failure of a predictive equation demonstrates that the equation cannot be proven within probability theory alone. To see why, recall the fact of elementary logic that a formula F does not follow from a theory T if there is a *model* of T that does not satisfy F, that is, if there is a case in which all the premises and sentences of T are true but F is false. Now, let T be probability theory and F the breeder's equation. A model of probability theory is any distribution that satisfies Kolmogorov's axioms. Hence our question is whether there is a probability distribution that dissatisfies the breeder's equation, that is, a distribution against

BOX 1 DERIVATION OF THE BREEDER'S EQUATION

Causal graph theory models causal relationships among variables, which in the present context include genes X, phenotype Z, fitness (offspring count) W, and so on. Presumably these variables have a causal influence on each other. Fitness W, for instance, is affected by an organism's phenotype Z, which develops from its genes X, which in turn are transmitted to offspring's genes X', which finally shape their phenotype Z'. Note that these causal relationships respectively correspond to phenotypic transition (T_2), development (T_1), reproduction (T_3, T_4), and offspring development (T_1) in Lewontin's scheme (Fig. 1.2). In the case of the breeder's equation, these causal processes are modeled as the causal graph below and parameterized by the structural equations on the right. For example, the topmost equation quantifies the causal influence of the phenotype Z on the fitness W, with the coefficient β measuring the strength of (linear) selection and E_W being an "error term" that summarizes other phenotypic or environmental factors that affect fitness. The parameterized model generates a probability distribution over the variables, from which the breeder's equation is obtained as its statistical function. In the same fashion, one can derive other transition functions such as the one-locus population genetics model (Eqn. 5.1) or Fisher's fundamental theorem of natural selection (Otsuka, 2014, 2016b).

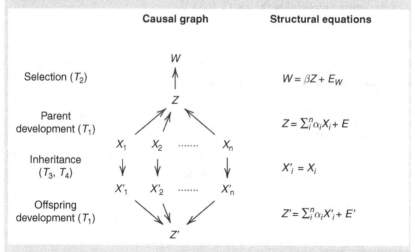

Causal graph   **Structural equations**

Selection (T_2) $W = \beta Z + E_W$

Parent development (T_1) $Z = \sum_i^n \alpha_i X_i + E$

Inheritance (T_3, T_4) $X'_i = X_i$

Offspring development (T_1) $Z' = \sum_i^n \alpha_i X'_i + E'$

Figure 5.1 The causal model for the breeder's equation

> BOX 2 THE AXIOMATIC BASIS OF CAUSAL GRAPH THEORY
>
> Formally speaking, a *causal model* is a pair of a causal graph and a probability distribution that satisfies several axioms. Causal graph theory studies the relationship between causality and probability by exploring the logical consequences or theorems from such axioms and auxiliary assumptions. Below are representative axioms and assumptions of the theory (for more detail see Spirtes et al., 1993; Pearl, 2000).
>
> - **Markov condition**: states that causally disconnected (called *d*-separated) variables are probabilistically independent.
> - **Faithfulness**: is the opposite of the Markov condition and states that probabilistically independent variables are causally disconnected.
> - **Trek rule** (Wright's method of path coefficient): calculates the covariance of two variables by summing up causal connections between them. This rule is used to obtain the breeder's equation from its causal model.

which the equality does not hold. But we have just seen such a counter model/ distribution: any distribution generated from the confounded causal model, although it satisfies all the axioms of probability theory as a *bona fide* distribution, nevertheless fails Eqn. 5.2. It thus follows that the breeder's equation is not a theorem of probability theory, Q. E. D.[16]

If anything, it is a theorem of causal graph theory – that's what the above discussion suggests. Causal graph theory is an axiomatic theory that allows us to deduce probabilistic or causal conclusions from a set of premises (Box 2). This deductive process does not differ qualitatively from those in other mathematical practices, say, when we solve an equation by applying the rules of elementary arithmetic. Indeed, causal graph theory does contain more elaborate definitions and axioms that afford it a broader expressive power and richer mathematical operations, but the essence remains the same: it deduces logical consequences of the axioms and premises. The causal model as depicted in Fig. 5.1 represents a set of such premises, and from these premises and the axioms of the theory follows the breeder's equation. This means that the equation is proven as a theorem within causal graph theory combined with the given premises.

---

[16] Some readers may find this result at odds with the oft-made claim that the breeder's equation and the population genetics model can be derived from the Price equation (Okasha, 2006; Frank, 1995; Queller, 2017). These derivations, however, rely on assumptions that are not themselves either axioms, theorems, or even statements of probability theory. The argument in this section is that these nonprobabilistic assumptions can be formally expressed by the language of causal graph theory.

Although this argument is far cry from a formal proof, it suffices as a rough sketch of the theoretical background of population genetics models. In particular, it suggests that a part of evolutionary genetics that studies genetic and phenotypic dynamics of Mendelian populations can be theoretically reconstructed, if not axiomatized, using the formal apparatus of causal graph theory.

One may wonder at this point how this appeal to causality sits with the aforementioned fact that celebrated principles of evolution such as the Price equation and Li's theorem are provable within probability theory (Section 1). As probability theorems they are applicable to any population regardless of its physical or causal substratum, and for this reason are often considered to be the most universal statements of evolution. But if so, why should we bother with causality? The problem with these probability statements is that they are purely descriptive. The Price equation and Li's theorem may well summarize a given evolutionary change in a concise equational form but never predict it *ante factum* (Section 3), and for this reason they cannot serve as the basis for ampliative reasoning about future evolutionary trajectories. This is contrasted with formulae like the one-locus population genetics model or the breeder's equation, whose aim is to predict an evolutionary response and trajectory based on the current population and environmental conditions. What we have just seen is that these predictive formulae do not follow from the axioms of probability, but require substantive assumptions about an evolving population, including those about its causal structure. Although these additional assumptions curtail the universality – for now the formulae apply only to populations that conform to the given causal models – they purchase the ampliative power to derive new conclusions about future evolutionary paths that are not contained in the premises. No causal assumption in, no evolutionary prediction out. This is the reason why the formal basis of evolutionary theory must comprise a theory that deals with not just probabilistic but also causal relationships.

5.2 The Empirical Nature of Evolutionary Theory

With this theoretical result in mind, let us go back to our original philosophical questions regarding the empirical and inductive nature of evolutionary theory. Our starting point was the tautology problem, the charge that evolutionary theory makes no empirical claims because its core principles are a priori mathematics. As I pointed out in Section 4, the implicit assumption underlying this charge is the strict distinction between analytic and synthetic statements, according to which a statement is either analytic or synthetic, and anything analytic/mathematical cannot be synthetic/empirical. This assumption, however, has been put in doubt ever since Quine's attack on logical empiricism. Even if we don't go that far, contemporary mathematics tells us that there are varieties of mathematical theories, with different formal languages and axioms. Hence the dichotomous demarcation of "tautology or not" is too blunt a knife to carve out the theoretical nature of evolutionary theory. We should at least start our inquiry by asking: if evolutionary theory is mathematical, by which mathematics?

Above I argued that the relevant mathematical theory to ground evolutionary induction is causal graph theory, within which the breeder's equation and other ampliative formulae of population genetics can be proven as theorems from certain sets of premises. But what does this imply for our philosophical investigation on the empirical nature of evolutionary theory? Causal graph theory is a mathematical theory to deal with the interrelationship between probability and causality – it uses the mathematical concept of the graph to represent causal relationships and deduce probabilistic consequences – and as such it has both a priori and a posteriori flavors. As an axiomatic theory, it builds upon probability theory and graph theory, with some additional axioms relating these two mathematical structures (Box 2). The theoretical results from this axiomatic framework look as deductive and a priori as any other theorems in mathematical theories. On the other hand, as a theory of causation it purports to express some general facts about the relationship between causal structures and statistical data, and these expressions are no less empirical and a posteriori than nomological statements in scientific theories, say Newton's laws of motion. Let's take the Markov condition, which asserts that two causally disconnected variables are statistically independent – that is, by contraposition, no genuine correlation without a causal connection. This is certainly not analytic or a priori, but makes a substantial claim about the world. And in fact the claim may not even be universally true, especially in the quantum realm, where the reported violation of the so-called Bell's inequality suggests that the spin state of a pair of particles can be correlated in a way not attributable to any causal connection between them – so there appears to be a correlation without causation, after all.

What are we to make of this dual nature of causal graph theory? One response would be to stick to the logical empiricist perspective and take these two aspects to form different components of the theory, namely logical calculus and bridge laws. Establishing logical relationships between a probability distribution and a graphical structure is one thing, interpreting these relationships to hold of actual statistical data and what we take to be causal structures of the world is another. The former is analytic while the latter is synthetic, and the theory *qua* study of causation combines these well-separated components. Let us grant all this for a moment; we still are in a much better position to answer the charge of tautology than before. This is because the charge applies only if evolutionary theory turns out to be *purely* mathematical or statistical, whatever that means, while we have already seen that causal graph theory is not of that kind. It is much less a piece of pure mathematics than a mathematically framed hypothesis about causal relationships, and as such epistemologically on a par with other mathematical scientific theories like, say, the theory of relativity. Moreover, the derivation of ampliative formulae of evolutionary changes requires particular sets of causal assumptions, which are far from a priori. Since these assumptions are essential ingredients of evolutionary theory, it should be regarded as at least as empirical as other formal sciences, even by the logical empiricist standard.

But those who follow Quine in rejecting the analytic/synthetic distinction together with the whole empiricist approach may prefer a more holistic picture. Recall that Quine viewed a scientific theory not as a combination of analytic and synthetic components, but as an integrated network with logical or theoretical statements around the core which impinges upon experience along the edges. Likewise, causal graph theory consists of various kinds of assumptions, some of which lie deep in discrete mathematics (graph theory) or probability theory while others – such as the Markov or faithfulness conditions – sit closer to the empirical edge, but without a sharp line that divides them into the revisable and the irrevisable. Above I noted that the axioms of causal graph theory are not immune to revision. The axioms of classical probability theory are usually regarded to be on safer ground, but not absolutely safe; indeed they may likewise be challenged by quantum or psychological phenomena, which, according to some theorists, are better captured by a nonclassical "quantum" probability theory (e.g., Pothos and Busemeyer, 2009). The difference between probabilistic and causal axioms in terms of their revisability, therefore, is not categorical but rather a matter of degree.

Surrounding this theoretical core of the belief network are descriptions of a particular causal system, to be specified in each application of the theory. In the case of the breeder's equation, these assumptions include the causal graph and structural equations of an evolving population as shown in Fig. 5.1, and other background conditions concerning the population size, structure, or the mating system that are not explicit in the figure. They form the outskirts of the theory and often become the first candidates for close scrutiny when predictions do not match with actual outcomes. Suppose, for instance, that an evolutionary response of a certain population turned out to be much smaller or greater than a prediction of the breeder's equation. Biologists facing such discrepancies are more likely to doubt the hypothesized causal graph or estimated parameters than to give up some principles of causal graph theory or probability theory. In this sense the causal assumptions of a particular population sit closer to the "empirical edge" of our belief system than do the theoretical axioms.

But not all empirical assumptions are equal. As was the case with theoretical axioms, biological assumptions come in different degrees of revisability in the event of the model's failure. One advantage of the Quinian holism is that it accounts for the difference in weight biologists assign to empirical assumptions. To see this, imagine a scientist who found that the actual evolutionary response of a population under study largely deviated from the prediction of the breeder's equation. As suggested above, a common reaction to such a failure is to suspect a confounding factor, that is, to examine the hypothesized causal relationship $Z{\rightarrow}W$. But this is certainly not the only way to tinker with the model, for one could also revise other parts of the graph or the structural equations to save the phenomenon. For example, one could question the Mendelian assumption that each gene has an equal chance to be inherited (so that $\mathbf{X}' \neq \mathbf{X}$);[17] or, contrary to the so-called "central dogma" of molecular biology, a parent's phenotype might affect its genotype $(Z{\rightarrow}X)$. We could stretch our imagination even further and pretend that, as in some bacteria, parents exchange their genetic material so that there are extra causal paths among parents. I am not suggesting that working biologists or breeders facing the failure of their prediction readily consider these possibilities. Mostly they don't, and that is because these hypotheses, though logically possible, contradict well-established tenets of evolutionary or other biological sciences, and revisions of such beliefs would have repercussions throughout the entire field. Hence, although these assumptions are all empirical and concern concrete living systems, they are more centrally located within the theory and less susceptible to falsification.

However, even those core tenets may be challenged, and sometimes abandoned, in cases of systematic failure or a mismatch of a model with reality. Revising the core doctrines requires an overall rearrangement or reevaluation of the belief network and often gives rise to a significant debate among scientists. A disagreement arises because not all researchers share the same system of beliefs, and as a consequence there is always leeway as to which assumption to drop or revise in the event of a model's failure. Recent discussions of *missing heritability* illustrate a case in point. In the past few decades the development of genome-wide association studies (GWAS) has enabled researchers to identify specific regions of DNA responsible for a given phenotype, but it has also been reported that the amount of genetic variation estimated using this technique tends to be much smaller than that obtained from the traditional heritability

---

[17] Such a biased transmission is known as *segregation distortion* and is observed in many species (Ridley, 2004, p. 294).

estimation. Lively discussions followed over potential sources of this discrepancy, including poor statistical power, epistatic effects, gene-environment interactions, epigenetic factors, transgenerational genetic effects, and so on (Manolio et al., 2009; Eichler et al., 2010; Bourrat and Qiaoying, 2017), some of which cut deep into the core tenets of canonical evolutionary theory. Each of these suggestions is motivated or underpinned by different theoretical or empirical grounds, such as predictions from population genetics theory or experimental studies on genetic as well as epigenetic factors. Biologists have thus pointed to different factors as the culprits of missing heritability based on their own respective "system of beliefs" – the view as to which assumptions form the core and which are on the edge of the whole theory of genetic variation and evolution.

To sum up, the holistic picture describes evolutionary theory as a network of beliefs with multiple layers, ranging from the core mathematical axioms of probability, causal graphs, statistics, and so on, to well-received doctrines of evolutionary or other biological sciences, to working hypotheses concerning particular populations under study. Its assertions and predictions about the biological world come only through these layers and confront the tribunal of empirical observations or experiments "as a corporate body" (Quine, 1951, p. 38). The goal of metascientific analysis according to this picture is to understand how this theoretical body functions as a whole and responds to possible conflicts with experience. In this endeavour, *why* mathematics can explain evolutionary phenomena is no longer in question. As far as evolutionary biology is concerned, mathematical axioms and models are part of our empirical theory to explain and predict evolutionary phenomena, and as such are epistemologically on a par with other causal or biological assumptions. What is more relevant and important is *how* they are embedded in the theoretical network and coordinate with other parts of the theory. The causal modeling framework addresses this question by combining probability theory with relevant causal/biological assumptions to yield quantitative predictions of evolutionary change. Granted, this is far from the entire picture of evolutionary theory, given that adequate studies of evolution must take into account a whole lot of other factors not readily represented or handled by causal models, such as population size, structure, and nonlinear evolutionary dynamics, to name a few. I hope, however, that the above discussion gives a partial sketch of the relationship between mathematical principles and other parts of evolutionary theory, as well as their role in the explanation of evolutionary phenomena.

5.3 The Generalizability of Evolutionary Explanations

Hence evolutionary theory is, after all, an empirical science. But this conclusion alone does not fully answer Smart's skepticism about the autonomy of the biological sciences. His challenge, recall, was a two-pronged one, of which the charge of tautology constituted only one horn. The other contention was that evolutionary theory, or biology in general, lacks general laws, and in this respect it is more like an application of the more basic sciences of physics or chemistry (biology as "radio engineering"). On the face of it, the Quinian holistic picture described above does not seem helpful in dismantling this reductionist threat: indeed, it might even make the matter worse, by pushing biological theories to the fringe around the harder core of fundamental sciences. This runs counter to the search for biological generality and antireductionism, which has been the primary concern of the philosophy of biology inherited from the pioneers of population genetics like Fisher and Haldane (Section 1). They expected mathematical models of population genetics to serve as the "guiding logical ideas" or laws of biological sciences, that distinguish evolutionary theory from descriptive natural history and at the same time establish it as an autonomous discipline irreducible to the physical sciences. A satisfactory metascientific account, therefore, must address this issue and explain the generalizability, if any, of evolutionary theory.

The modus operandi of the traditional philosophical literature to tackle this problem was to resort to the notion of supervenience (Sections 2 and 3). Since the key parameters of mathematical models of evolution, such as the fitness values or other statistical properties of a population, can be realized by various physical substrata, the same model is applicable to different populations or situations regardless of phenotypic or environmental idiosyncrasies. In other words, supervenience allows us to abstract away unnecessary details and reveal the essential patterns. These macroscopic patterns are not visible in the reductive or physicalistic description but emerge only in the bird's-eye view of evolutionary theory, or so it was argued.

The above discussion of causal modeling, however, suggests that the supervenience of a model's parameters does not alone warrant its applicability – how they arise from the causal nexus also matters. One may well identify and measure relevant variables such as fitness or phenotype, correctly estimate their expected values or correlation coefficients, and plug these estimates into the breeder's equation, but still fail to predict evolutionary response if the requisite causal assumptions are not met, as we have seen above in the case of confounding. Successful applications of evolutionary models are contingent on the causal structure of target populations. This conclusion might appear to compromise the generality of evolutionary theory and lean toward reductionism. Causation is often understood as a concrete relationship between physical objects, and there are myriad ways in which causal interactions take place among organisms and environment. Organisms' births, lives, and deaths are influenced by varieties of contingent factors that differ from time to time and place to place. If evolutionary models were based on such causal details, wouldn't studies of evolutionary change lose general applicability and be reduced to case-by-case inspections of particular causal mechanisms?

What underlies this worry is the mechanistic understanding of causality as specific physical processes or configurations that mediate interactions among concrete individuals (e.g., Salmon, 1984; Machamer et al., 2000). This, however, differs markedly from the notion of causality understood in the causal modeling literature that is required for the derivation of the evolutionary formulae above. A causal model is a description not of concrete physical mechanisms but rather of general counterfactual relationships between causes and effects, where graphs and structural equations specify whether and how an effect variable would change upon a hypothetical intervention on its cause variable(s) (Woodward, 2003). Such a counterfactual relationship may be instantiated by a variety of mechanisms – it is for this reason that we can anticipate what would happen when we press the brake pedal of a car without possessing the slightest idea about the specific mechanism linking the pedal and wheels, which may indeed differ across automobile companies. Thus causal relationships as encoded by causal models are still abstract, counterfactual patterns between variables that supervene on and are realized by a variety of physical mechanisms.

Likewise, causal models underlying evolutionary formulae specify general patterns that supervene on numerically distinct organisms, and this opens up the possibility of extrapolating the same evolutionary models to distinct populations or over multiple generations. These models concern only the structural features of causal relationships, that is, how variables are related to each other, without regard to the physiological nature of these variables. Hence the breeder's equation is applicable to populations of any organism whose phenotypes, genotypes, and environmental factors instantiate the general causal pattern as described by Fig. 5.1, regardless of their physical construction or the physiological mechanisms connecting them. Similarly, population genetics formulae can be recursively applied over multiple generations as long as the causal effects of the genes remain the same despite changes in gene frequencies. Conversely, if evolution or environmental fluctuations alter not just a probabilistic distribution but also the nature of causal connections among the variables, as in the cases of gene-by-gene or gene-by-environment interactions, the application of the same evolutionary model is no longer justified. It is thus the isomorphism of causal structure, and not just the supervenience of fitness values, that underlies extrapolation of evolutionary models, both diachronically and synchronically.

This shift of the focus from the fitness variable to the underlying causal structures has important implications for our understanding of the inductive nature of evolutionary theory. In Section 1 we saw that population genetics saved Darwin's theory of evolution by providing quantitative means to predict and calculate evolutionary dynamics. But on what grounds is such inductive reasoning justified? As Hume pointed out, inductive reasoning assumes the uniformity of nature, that is, some sort of invariance across numerically distinct objects, phenomena, or situations about which we make an inference. Hence if we wish to recursively apply the same population genetics formula for several generations as in Fig. 1.3 or use the breeder's equation for different populations, we must assume that these generations or populations retain or share some common feature despite apparent distributional, biological, or environmental changes or differences. Supervenience was expected to serve as such a uniform nature, a metaphysical anchor of evolutionary induction in the fundamental variability of the biological world (Section 2). The anchor, however, doesn't quite reach the floor. The agreement in or supervenience of fitness, phenotypic, or genetic distribution alone does not authorize an extrapolation of evolutionary formulae from one population to another; they must also share the same causal structure. Conversely, insofar as different populations instantiate the same causal model, their evolutionary dynamics can be captured by the same formula, provided that the other conditions not explicitly specified by the model, such as migration rate or population size, are equal. The causal structure thus serves as the basic unit of induction, the underlying "uniformity of nature" that authorizes the use of population genetics models to study evolutionary trajectories beyond observation.

The assumption of uniformity was crucial for Darwin's original argument, where he tried to convince his readers of the power of natural selection through an extrapolation of what skilled breeders and horticulturists can achieve by artificial selection to what Nature could achieve in a much longer time scale:

> Slow though the process of selection may be, if feeble man can do much by his powers of artificial selection, I can see no limit to the amount of change, to the beauty and infinite complexity of the coadaptations between all organic beings, one with another and with their physical conditions of life, which may be effected in the long course of time by nature's power of selection. (Darwin, 2003)

The argument stands up only if one can reasonably assume the conditions favorable for successful breeding, such as a continuous supply of variations, diligent check of the unfit, and heritability, continue to hold during "the long course of time" in which Nature carries through her job of cumulative selection. But in addition to these empirical premises, the veracity of the conclusion also depends on the logical force of the inductive reasoning itself, that is, whether the continuous fulfillment of these conditions really leads to the significant differences we observe between species, genera, and other higher taxa, and if so, how fast. In this respect, Darwin's argument wanted not only a verification or elucidation of its inductive assumptions – "the uniform nature" – but also the precise formulation of these assumptions that is requisite to assessing its argumentative force. The former was partly provided by Mendelian genetics, which furnished the basic principles of inheritance unknown to Darwin, while the latter was addressed by the mathematical formalization of evolutionary processes in population genetics. By incorporating Mendelian genetics, population genetics provided a formal framework to represent the empirical conditions necessary for adaptive evolution and forged Darwin's ingenious but qualitative reasoning into quantitative predictions of evolutionary changes. It is through this mathematization of the "uniformity of nature" that population genetics has enabled a precise formulation of natural selection and an evaluation of its capacity to shape organisms or species in comparison to other evolutionary factors.

Since many such inductive assumptions, including those concerning the nature of inheritance or selection, are of a causal nature, the discussion in the present section naturally centers around the mathematical formulation of the causal structures of evolving populations at the expense of other kinds of assumptions about, say, mutation or migration rate, population size or structure, etc. This choice of focus by no means implies the relative insignificance of the latter. The relative importance can be evaluated only with a model that incorporates both of them, and that was precisely one of the main points of using a quantitative framework. Moreover, the evolutionary dynamics treated here are arguably the simplest ones, ignoring stochastic effects or all the complex interactions that can take place among individuals, between populations, or with the environment. Dealing with these complex situations would require alternative mathematical frameworks such as diffusion theory, nonlinear dynamics, game theory, and so on. But whatever formal apparatus is employed, mathematical modeling serves for evolutionary theorizing only insofar as it correctly captures invariant features of an evolving population that theoretically anchor inductive inferences.[18] Such quantitative formulation of "the uniformity of

---

[18] In nonlinear cases like dominance, epistasis, frequency-dependent selection, etc., evolutionary dynamics depend on the population frequencies. But the model must still capture some invariant, presumably causal, features of these interactions in order to track the dynamics. The same applies to optimality models, which are concerned primarily with stable equilibria rather than

nature" is the single most important role of mathematics in evolutionary theory, which has contributed to forging the study of evolution into a precise and exact scientific theory with its own inductive principles.

6 Conclusion

This Element has explored the major philosophical issues concerning the role and use of mathematics in evolutionary biology, with a particular emphasis on their implications for our metascientific understanding of the structure of evolutionary theory. In closing our short journey, let us return to our original question: why mathematize?

First, mathematical models serve as proof-of-concept, making precise the logic and assumptions of a verbally stated hypothesis to demonstrate that the argument carries through as intended (Haldane, 1964; Servedio et al., 2014). As Haldane (1964) observed, had Newton provided only a verbal formulation of his theory of universal gravitation, very few would have believed that the attractions between the sun and planets maintain the planets' stable elliptic orbits without collapsing them into one massive body. It was the power of mathematical formulation and calculation that convinced his contemporaries of the veracity of his revolutionary hypothesis. In the same vein, by rendering Darwin's verbal argument in his *Origin* into precise and quantitative formulae, population genetics in the early twentieth century demonstrated that natural selection can effectively produce the visible and significant evolutionary changes Darwin hypothesized (Section 1).

Second, mathematization, or any kind of formalization in general, introduces the form/content distinction that enables scientists to study and explore their research subject as a formal system without and beyond concrete physical or biological phenomena (Griesemer, 2012). An excellent example of this is the schematic representation of evolution in the phenotypic and genotypic state spaces (Fig. 1.2). The formal rendering not only allowed researchers to analyze evolutionary processes independently from the idiosyncratic history and environment of each biological population, but also promoted further theorizing and new research questions regarding the nature of transition functions and the abstract spaces.

Mathematization thus serves for both the confirmation and the construction of a scientific theory. Either way, the key to its success consists in distilling the core assumptions and logic of the theory while abstracting inessential

---

step-by-step changes. These equilibria are concluded as outcomes of multiple turns of adaptive evolution, during which (at least some of) the model's parameters must stay invariant.

idiosyncrasies or "contents." And since the kinds of expressible assumptions depend on the language used, it highlights the importance of choosing an appropriate mathematical framework to describe the "form." This choice must be made in accordance with the goal of scientific theorizing, which is an inherently pragmatic activity aimed at a particular epistemic purpose, as Griesemer (2012) observes. Evolutionary theory is multifaceted and comprises various epistemic goals including phylogenetic inference, reconstruction of extinct life forms, and prediction of evolutionary trajectories, to name just a few. The diversity in evolutionary investigations necessitates different logic, assumptions, and mathematical frameworks. The use of causal graph theory in the previous section as a basis for population genetics models should be understood in this pragmatic context. That choice was made especially for the purpose of making sense of the *ampliative nature* of evolutionary transition functions: that is, its aim was to formalize the inductive reasoning of evolutionary change. Predicting evolutionary responses, however, hardly exhausts the theoretical goals even of evolutionary genetics. For example, one may just be interested in a descriptive summary of evolving populations, and for that purpose probability theory or even elementary arithmetic may suffice (e.g., Baker, 2005). In contrast, understanding the evolutionary significance of stochastic factors, interpopulation competitions, or organism-to-organism interactions would require more elaborate mathematical frameworks such as diffusion theory, dynamic systems theory, or game theory. From a pragmatic perspective, mathematics is a tool to solve given problems, and just as mathematicians create new mathematical tools or theories, theoretical evolutionary biologists build mathematical models that best serve to study the evolutionary phenomena in question.

This Element suggested that we understand these roles of mathematics in evolutionary theory from a Quinian perspective. Instead of drawing a sharp distinction between mathematical modeling and empirical theorizing, the Quinian picture describes mathematics as an integral part of evolutionary theory, epistemologically on a par with its causal and biological assumptions. This implies that what we call evolutionary theory is not to be delineated by some "fundamental principles" or necessary and sufficient conditions, but rather by the configuration and connections of a belief network that functions as a whole to contribute to our understanding of evolutionary phenomena. The nature of evolutionary theory, then, is best understood by elucidating the logical connections and coordinations of this belief network. How are different parts of the theory related to each other, and how do they function together to interact with empirical data? There are various ways to tackle these problems, but here let me sketch just a few open questions before we close.

The first task is to delineate various branches of evolutionary biology and their logical connections within the general picture of evolutionary theory. This Element focused on a particular task of evolutionary genetics, namely the prediction of short-term evolutionary responses. Arguably this is just a fraction of a larger network of evolutionary theory that covers diverse subjects such as ecology, development, phylogenetics, paleontology, and so on. If these fields represent distinct regions of the entire belief network of evolutionary theory, as the Quinian picture suggests, it is pertinent to ask about the nature of the logical or theoretical relationships that unite (or disunite) them under the name of evolutionary theory. There can hardly be just a single correct approach to tackle this big question, but I believe that attending to the formalism sheds light on the nature of the intertheoretical connections; for formalization, as we noted above, allows for a characterization of theories independent of their subject matters and helps to unravel commonalities beyond heterogeneity. In addition, even though the mathematical apparatus relevant to a given subfield may and should vary depending on its epistemic goal, mathematical theories themselves are not unrelated but often show logical relationships to each other. Take, for instance, the concept of species, which is arguably one of the most central but also notoriously polysemous concepts with dozens of different definitions in circulation with varying theoretical purposes. Formalizing the theoretical role played by the species concept in each context may help us to relate its different meanings and usages (Otsuka, forthcoming b). In this way, specifying and comparing the formal concepts and assumptions requisite for various agendas of evolutionary studies can be expected to clarify their mutual relationships and the functioning of the theory as a whole.

The second question is how the theoretical body of evolutionary theory relates to empirical phenomena and "faces the tribunal of observed data." The discussion in the previous section focused exclusively on the internal construction of population genetics and the identification of its theoretical as well as empirical assumptions. But in order to apply a model to real populations, its assumptions must be tested and its parameters estimated with observational or experimental data. How is this achieved? Quine described confirmation and estimation not as simple one-to-one correspondences between theoretical components and pieces of experience, but as holistic processes that depend on the entire belief network. Suppes (1962, 1967) offered a more detailed account of this complex relationship, wherein a formal theory relates to empirical data through a hierarchy of different models including models of theory, experiment, data, and measurement. The importance of verifying the assumptions of measurement and statistical analyses has recently been stressed by several authors (Houle et al., 2011; Huttegger and Mitteroecker, 2011). In addition, the discussion in the previous section further suggests that *causal discovery and testing* (Spirtes et al., 1993; Pearl, 2000; Imbens and Rubin, 2015) is an essential part of the confirmation process that relates population genetics theory to experimental or observational data. These new techniques have gradually been introduced to the population genetics literature (Shipley, 2000; Hadfield, 2008; Valente et al., 2010; Otsuka, 2016c), but their implications for the general methodology and practice of evolutionary biology remain to be explored.

Last but not least, philosophical reflections on formalism in evolutionary theory will have ontological implications that influence our view about what there are to evolve. In the evolutionary literature the canonical ontology has been given by Ernst Mayr's *population thinking*, according to which biological entities, either individual organisms or their parts, are fundamentally hetero-geneous and have no intrinsic commonality with each other (Mayr, 1975). In Section 2 we saw that this led to the ostracisation of *natural kinds* from the biological realm, which had traditionally served as fundamental building blocks for scientific theories in physics and chemistry. But can biology really get along without natural kinds? The primary role of natural kinds is to support inductive reasoning by embodying the Humean uniformity that justifies an extrapolation from the observed to the unobserved, the explanans to the explanandum (Boyd, 1991). And as we have seen, the role of mathematics in evolutionary genetics is to articulate this uniformity of nature in a formal and precise language. If so, the application of mathematics to evolutionary studies must assume, and its success in predicting population dynamics to some extent supports, the existence of some form or another of biological kinds (Otsuka, 2017, forthcoming a). The recent work by Benjamin Jantzen (2015, 2017) elaborates this line of reasoning and formulates evolutionary kinds and units according to the *symmetry* of evolutionary dynamics. The idea, in short, is that evolutionary kinds are defined in terms of the sameness of evolutionary dynamics, which in turn is determined by invariance against a group of possible interventions (which form symmetry transformations). Given the pivotal role of the symmetry principle in mediating fundamental equations and particles in physics (van Fraassen, 1989; Brading and Castellani, 2003), one can expect that its application to theoretical studies of evolution may also bring fruitful results (see also Frank, 2015).

Each of these questions hinges upon the role of mathematical and formal reasoning in evolutionary theory. Thus mathematics stands at the center of philosophical investigations on the theoretical nature, methodology, and ontol-ogy of evolutionary theory, and for this reason it will remain an important subject area and rich resource in the philosophy of evolutionary biology.

References

Baker, A. (2005). Are There Genuine Mathematical Explanations of Physical Phenomena? *Mind*, 114(454):223–238.

Beatty, J. (1980). Optimal-Design Models and the Strategy of Model Building in Evolutionary Biology. 47(4):532–561.

Beatty, J. (1981). What's Wrong with the Received View of Evolutionary Theory? *PSA: Proceedings of the Biennial Meeting of the Philosophy of Science Association*, Vol. 1980, 2:397–426.

Beatty, J. (1995). The Evolutionary Contingency Thesis. In Wolters, G. and Lennox, J. G., editors, *Concepts, Theories, and Rationality in the Biological Sciences*, pp. 45–81. University of Pittsburgh Press.

Bouchard, F. and Rosenberg, A. (2005). Matthen and Ariew's Obituary for Fitness: Reports of Its Death Have Been Greatly Exaggerated. *Biology & Philosophy*, 20(2–3):343–353.

Bourrat, P. and Qiaoying, L. (2017). Dissolving the Missing Heritability Problem. *Philosophy of Science*, (84):1055–1067.

Bowler, P. J. (1983). *The Eclipse of Darwinism: Anti-Darwinian Evolution Theories in the Decades around 1900*. John Hopkins University Press, Baltimore, MD.

Boyd, R. N. (1991). Realism, Anti-foundationalism and the Enthusiasm for Natural Kinds. *Philosophical Studies*, 61(1–2):127–148.

Brading, K. and Castellani, E. (2003). *Symmetries in Physics: Philosophical Reflections*. Cambridge University Press, Cambridge, UK.

Brandon, R. N. (1981). A Structural Description of Evolutionary Theory. In *PSA: Proceedings of the Biennial Meeting of the Philosophy of Science Association*, Vol. 1980, pp. 427–439.

Brandon, R. N. and Beatty, J. (1984). The Propensity Interpretation of 'Fitness' – No Interpretation Is No Substitute. *Philosophy of Science*, 51(2):342–347.

Carnap, R. (1966). *An Introduction to the Philosophy of Science*. Dover Publications, Inc., Mineola, NY.

Darwin, C. (2003). *On the Origin of Species: A Facsimile of the First Edition*. Harvard University Press, Cambridge, MA.

Eichler, E. E., Flint, J., Gibson, G., Kong, A., Leal, S. M., Moore, J. H., and Nadeau, J. H. (2010). Missing Heritability and Strategies for Finding the Underlying Causes of Complex Disease. *Nature Reviews Genetics*, 11(6):446–450.

Endler, J. A. (1986). *Natural Selection in the Wild*. Princeton University Press, Princeton, NJ.

Fisher, R. A. (1918). The Correlation between Relatives on the Supposition of Mendelian Inheritance. *Transactions of the Royal Society of Edinburgh*, 51:399–433.

Fisher, R. A. (1930). *The Genetical Theory of Natural Selection*. Oxford University Press, Oxford.

Fodor, J. A. (1974). Special Sciences (or: The Disunity of Science as a Working Hypothesis). *Synthese*, 28(2):97–115.

Frank, S. A. (1995). George Price's Contributions to Evolutionary Genetics. *Journal of Theoretical Biology*, 175:373–388.

Frank, S. A. (1997). The Price Equation, Fisher's Fundamental Theorem, Kin Selection, and Causal Analysis. *Evolution*, 51(6):1712–1729.

Frank, S. A. (2012). Natural Selection. IV. The Price Equation. *Journal of Evolutionary Biology*, 25(6):1002–1019.

Frank, S. A. (2015). D'Alembert's Direct and Inertial Forces Acting on Populations: The Price Equation and the Fundamental Theorem of Natural Selection. *Entropy*, 17(10):7087–7100.

Friedman, M. (1974). Explanation and Scientific Understanding. *The Journal of Philosophy*, 71(1):5–19.

Goodman, N. (1955). *Fact, Fiction, and Forecast*. Harvard University Press, Cambridge, MA.

Gould, S. J. (1976). Darwin's Untimely Burial. *Natural History*, 85(8):24–30.

Grafen, A. (2014). The Formal Darwinism Project in Outline. *Biology & Philosophy*, 29(2):155–174.

Griesemer, J. R. (2012). Formalization and the Meaning of "Theory" in the Inexact Biological Sciences. *Biological Theory*, 7(4):298–310.

Hadfield, J. D. (2008). Estimating Evolutionary Parameters When Viability Selection Is Operating. *Proceedings of the Royal Society B: Biological Sciences*, 275(1635):723–734.

Haldane, J. B. S. (1964). A Defense of Beanbag Genetics. *Perspectives in Biology and Medicine*, 7(3):434–359.

Haldane, J. B. S. (1931). *The Philosophical Basis of Biology*. Hodder and Stoughton Limited, London, UK.

Houle, D., Hansen, T. F., Pélabon, C., and Wagner, G. P. (2011). Measurement and Meaning in Biology. *The Quarterly Review of Biology*, 86(1):3–34.

Hull, D. L. (1974). *Philosophy of Biological Science*. Prentice-Hall, Englewood Cliffs, NJ.

Huttegger, S. M. and Mitteroecker, P. (2011). Invariance and Meaningfulness in Phenotype Spaces. *Evolutionary Biology*, 38(3):335–351.

Imbens, G. W. and Rubin, D. (2015). *Causal Inference in Statistics, Social, and Biomedical Sciences*. Cambridge University Press, New York, NY.

Jantzen, B. C. (2015). Projection, Symmetry, and Natural Kinds. *Synthese*, 192(11):3617–3646.

Jantzen, B. C. (2019). Kinds of Process and the Levels of Selection. *Synthese*, 196(6):2407–2433.

Kim, J. (1992). Multiple Realization and the Metaphysics of Reduction. *Philosophy and Phenomenological Research*, 52(1):1–26.

Kitcher, P. (1981). Explanatory Unification. *Philosophy of Science*, 48:507–531.

Lewontin, R. C. (1970). The Units of Selection. *Annual Review of Ecology and Systematics*, 1:1–18.

Lewontin, R. C. (1974). *The Genetic Basis of Evolutionary Change*. Columbia University Press, New York, NY.

Li, C. C. (1955). *Population Genetics*. The University of Chicago Press, Chicago, IL.

Lloyd, E. A. (1988). *The Structure and Confirmation of Evolutionary Theory*. Princeton University Press, Princeton, NJ.

Luque, V. J. (2017). One Equation to Rule Them All: A Philosophical Analysis of the Price Equation. *Biology & Philosophy*, 32(1):97–125.

Machamer, P., Darden, L., and Craver, C. F. (2000). Thinking about Mechanisms. *Philosophy of Science*, 67(1):1–25.

Magalhães, J. C. M. and Krause, D. (2001). Suppes Predicate for Genetics and Natural Selection. *Journal of Theoretical Biology*, 209(2):141–153.

Manolio, T. A., Collins, F. S., Cox, N. J., Goldstein, D. B., Hindorff, L. A., Hunter, D. J., McCarthy, M. I., Ramos, E. M., Cardon, L. R., Chakravarti, A., Cho, J. H., Guttmacher, A. E., Kong, A., Kruglyak, L., Mardis, E., Rotimi, C. N., Slatkin, M., Valle, D., Whittemore, A. S., Boehnke, M., Clark, A. G., Eichler, E. E., Gibson, G., Haines, J. L., Mackay, T. F. C., McCarroll, S. A., and Visscher, P. M. (2009). Finding the Missing Heritability of Complex Diseases. *Nature*, 461(7265):747–753.

Matthen, M. (2009). Drift and "Statistically Abstractive Explanation." *Philosophy of Science*, 76:464–487.

Matthen, M. and Ariew, A. (2002). Two Ways of Thinking about Fitness and Natural Selection. *Journal of Philosophy*, 99(2):55–83.

Matthen, M. and Ariew, A. (2005). How to Understand Casual Relations in Natural Selection: Reply to Rosenberg and Bouchard. *Biology & Philosophy*, 20(2–3):355–364.

Matthen, M. and Ariew, A. (2009). Selection and Causation. *Philosophy of Science*, 76(2):201–224.

Mayr, E. (1975). *Evolution and the Diversity of Life: Selected Essays*. Harvard University Press, Cambridge, MA.

Mayr, E. (1982). *The Growth of Biological Thought. Diversity, Evolution, and Inheritance.* The Belknap Press of Harvard University Press, Cambridge, MA.

Mckinsey, J. C. C., Suppes, P., and Sugar, A. C. (1953). Axiomatic Foundations of Classical Particle Mechanics. *Journal of Rational Mechanics and Analysis*, 2(2):253–272.

Mills, S. K. and Beatty, J. (1979). The Propensity Interpretation of Fitness. *Philosophy of Science*, 46:263–286.

Millstein, R. L. (2006). Natural Selection as a Population-Level Causal Process. *The British Journal for the Philosophy of Science*, 57(4):627–653.

Millstein, R. L., Skipper, R. A. J., and Dietrich, M. R. (2009). (Mis)interpreting Mathematical Models: Drift As a Physical Process. *Philosophy, Theory, and Practice in Biology*, 1:e002.

Morrison, M. (2000). *Unifying Scientific Theories. Physical Concepts and Mathematical Structures.* Cambridge University Press, Cambridge, UK.

Morrissey, M. B., Kruuk, L. E. B., and Wilson, A. J. (2010). The Danger of Applying the Breeder's Equation in Observational Studies of Natural Populations. *Journal of Evolutionary Biology*, 23(11):2277–2288.

Okasha, S. (2006). *Evolution and the Levels of Selection*. Oxford University Press, Oxford.

Otsuka, J. (2014). The Causal Structure of Evolutionary Theory. PhD diss., Indiana University.

Otsuka, J. (2016a). A Critical Review of the Statisticalist Debate. *Biology & Philosophy*, 31(4):459–482.

Otsuka, J. (2016b). Causal Foundations of Evolutionary Genetics. *The British Journal for the Philosophy of Science*, 67:247–269.

Otsuka, J. (2016c). Discovering Phenotypic Causal Structure from Nonexperimental Data. *Journal of Evolutionary Biology*, 29(6):1268–1277.

Otsuka, J. (2017). The Causal Homology Concept. *Philosophy of Science*, 84(5):1128–1139.

Otsuka, J. (forthcoming a). Ontology, Causality, and Methodology of Evolutionary Research Programs. In Uller, T. and Laland, K. N., editors, *Vienna Series in Theoretical Biology Evolutionary Causation*. The Massachusetts Institute of Technology Press.

Otsuka, J. (forthcoming b). Species as Models. *Philosophy of Science*, 86(5).

Pearl, J. (2000). *Causality: Models, Reasoning, and Inference*. Cambridge University Press, New York, NY.

Pence, C. H. and Ramsey, G. (2013). A New Foundation for the Propensity Interpretation of Fitness. *The British Journal for the Philosophy of Science*, 64(4):851–881.

Pigliucci, M. and Kaplan, J. (2006). *Making Sense of Evolution: The Conceptual Foundations of Evolutionary Biology*. University of Chicago Press, Chicago, IL.

Popper, K. R. (1974). Autobiography of Karl Popper. In Schlipp, P. A., editor, *The Philosophy of Karl Popper*. La Salle, IL.

Pothos, E. M. and Busemeyer, J. R. (2009). A Quantum Probability Explanation for Violations of 'Rational' Decision Theory. *Proceedings of the Royal Society B: Biological Sciences*, 276(1665):2171–2178.

Price, G. R. (1970). Selection and Covariance. *Nature*, 227:520–521.

Provine, W. B. (2001). *The Origins of Theoretical Population Genetics: With a New Afterword*. University of Chicago Press, Chicago, IL.

Punnett, R. C. (1915). *Mimicry in Butterflies*. Cambridge University Press, Cambridge, UK.

Putnam, H. (1975). Philosophy and our Mental Life. In *Mind, Language, and Reality*, pp. 291–303. Cambridge University Press, Cambridge, UK.

Queller, D. C. (2017). Fundamental Theorems of Evolution. *The American Naturalist*, 189(4):345–353.

Quine, W. V. O. (1951). Two Dogmas of Empiricism. *The Philosophical Review*, 60(1):20–43.

Ramsey, G. (2013). Can Fitness Differences Be a Cause of Evolution? *Philosophy, Theory, and Practice in Biology*, 5:e401.

Ramsey, G. (2016). The Causal Structure of Evolutionary Theory. *Australasian Journal of Philosophy*, 94(3):421–434.

Ramsey, G. and Brandon, R. N. (2007). What's Wrong with the Emergentist Statistical Interpretation of Natural Selection and Random Drift? In Hull, D. L. and Ruse, M., editors, *The Cambridge Companion to the Philosophy of Biology*, pp. 66–84. Cambridge University Press, Cambridge, UK.

Rausher, M. D. (1992). The Measurement of Selection on Quantitative Traits: Biases Due to Environmental Covariances between Traits and Fitness. *Evolution*, 46(3):616–626.

Reisman, K. and Forber, P. (2004). Manipulation and the Causes of Evolution. 72:1113–1123.

Rice, S. H. (2004). *Evolutionary Theory: Mathematical and Conceptual Foundations*. Sinauer Associates, Sunderland, MA.

Ridley, M. (2004). *Evolution*. Blackwell Publishing, Malden, MA, 3rd edition.

Robertson, A. (1966). A Mathematical Model of the Culling Process in Dairy Cattle. *Animal Production*, 8(1):95–108.

Rosenberg, A. (1982). On the Propensity Definition of Fitness. *Philosophy of Science*, 49:268–273.

Rosenberg, A. (1985). *The Structure of Biological Science*. Cambridge University Press, Cambridge.

Ruse, M. (1973). *The Philosophy of Biology*. Hutchinson University Library, London.

Ryle, G. (1949). *The Concept of Mind*. Hutchinson's University Library, London.

Salmon, W. C. (1984). *Scientific Explanation and the Causal Structure of the World*. Princeton University Press, Princeton, NJ.

Servedio, M. R., Brandvain, Y., Dhole, S., Fitzpatrick, C. L., Goldberg, E. E., Stern, C. A., Van Cleve, J., and Yeh, D. J. (2014). Not Just a Theory – The Utility of Mathematical Models in Evolutionary Biology. *PLoS Biology*, 12(12):e1002017–5.

Shipley, B. (2000). *Cause and Correlation in Biology: A User's Guide to Path Analysis, Structural Equations and Causal Inference*. Cambridge University Press, Cambridge, UK.

Smart, J. J. C. (1959). Can Biology Be an Exact Science? *Synthese*, 11(4):359–368.

Smocovitis, V. B. (1996). *Unifying Biology: The Evolutionary Synthesis and Evolutionary Biology*. Princeton University Press, Princeton, NJ.

Sober, E. (1984). *The Nature of Selection: Evolutionary Theory in Philosophical Focus*. University of Chicago Press, Chicago, IL.

Sober, E. (1993). *Philosophy of Biology*. Westview Press, Boulder.

Sober, E. (1997). Two Outbreaks of Lawlessness in Recent Philosophy of Biology. *Philosophy of Science*, 64(Proceedings):S458–S467.

Spirtes, P., Glymour, C., and Scheines, R. (1993). *Causation, Prediction, and Search*. The Massachusetts Institute of Technology Press, Cambridge, MA.

Suppes, P. (1962). Models of data. In *Logic, Methodology, and Philosophy of Science: Proceedings of the 1960 International Congress*, pp. 252–261.

Suppes, P. (1967). What Is a Scientific Theory? In *Philosophy of Science Today*, pp. 55–67. Basic Books, Inc., New York, NY.

Suppes, P. (2002). *Representation and Invariance of Scientific Structures*. CSLI Publication, Stanford, CA.

Tabery, J. (2014). *Beyond Versus: The Struggle to Understand the Interaction of Nature and Nurture*. The Massachusetts Institute of Technology Press, Cambridge, MA.

Valente, B. D., Rosa, G. J. M., De los Campos, G., Gianola, D., and Silva, M. A. (2010). Searching for Recursive Causal Structures in Multivariate Quantitative Genetics Mixed Models. *Genetics*, 185(2):633–644.

van Fraassen, B. C. (1989). *Laws and Symmetry*. Oxford University Press, Oxford, UK.

van Veelen, M. (2005). On the Use of the Price Equation. *Journal of Theoretical Biology*, 237(4):412–426.

van Veelen, M., Garca, J., Sabelis, M. W., and Egas, M. (2012). Group Selection and Inclusive Fitness Are Not Equivalent; the Price Equation vs. Models and Statistics. *Journal of Theoretical Biology*, 299:64–80.

Walsh, D. M., Ariew, A., and Matthen, M. (2017). Four Pillars of Statisticalism. *Philosophy & Theory in Biology*, 9(20171201).

Walsh, D. M., Lewins, T., and Ariew, A. (2002). The Trials of Life: Natural Selection and Random Drift. *Philosophy of Science*, 69(3):429–446.

Wigner, E. (1960). The Unreasonable Effectiveness of Mathematics in the Natural Sciences. *Communications in Pure and Applied Mathematics*, 13(1):1–14.

Williams, M. B. (1970). Deducing the Consequences of Evolution: A Mathematical Model. *Journal of Theoretical Biology*, 29:343–385.

Woodward, J. (2003). *Making Things Happen*. Oxford University Press, New York, NY.

Wright, S. (1934). The Method of Path Coefficients. *The Annals of Mathematical Statistics*, 5:161–215.

Acknowledgments

I thank Victor Luque, Samir Okasha, two anonymous reviewers, and the students in the philosophy of biology seminar taught at Kyoto University in 2017 for their comments and discussions. Grant Ramsey and Samuel Mortimer offered helpful suggestions and spotted numerous clerical errors in the draft, which I greatly appreciate. All remaining errors are my own. This work was supported by JSPS KAKENHI Grant 16K16335.

Cambridge Elements

Philosophy of Biology

Grant Ramsey
KU Leuven

Grant Ramsey is a BOFZAP Research Professor at the Institute of Philosophy, KU Leuven, Belgium. His work centers on philosophical problems at the foundation of evolutionary biology. He has been awarded the Popper Prize twice for his work in this area. He also publishes in the philosophy of animal behavior, human nature, and the moral emotions. He runs the Ramsey Lab (theramseylab.org), a highly collaborative research group focused on issues in the philosophy of the life sciences.

Michael Ruse
Florida State University

Michael Ruse is the Lucyle T. Werkmeister Professor of Philosophy and the Director of the Program in the History and Philosophy of Science at Florida State University. He is Professor Emeritus at the University of Guelph, in Ontario, Canada. He is a former Guggenheim fellow and Gifford lecturer. He is the author or editor of over sixty books, most recently *Darwinism as Religion: What Literature Tells Us about Evolution; On Purpose; The Problem of War: Darwinism, Christianity, and their Battle to Understand Human Conflict;* and *A Meaning to Life.*

About the Series

This Cambridge Elements series provides concise and structured introductions to all of the central topics in the philosophy of biology. Contributors to the series are cutting-edge researchers who offer balanced, comprehensive coverage of multiple perspectives, while also developing new ideas and arguments from a unique viewpoint.

Cambridge Elements ☰

Philosophy of Biology

Elements in the Series

The Biology of Art
Richard A. Richards

The Darwinian Revolution
Michael Ruse

Ecological Models
Jay Odenbaugh

Mechanisms in Molecular Biology
Tudor M. Baetu

The Role of Mathematics in Evolutionary Theory
Jun Otsuka

A full series listing is available at: www.cambridge.org/EPBY

Printed in the United States
By Bookmasters